From Poison Arrows to Prozac

STANLEY FELDMAN

From Poison Arrows to Prozac

HOW DEADLY TOXINS CHANGED OUR LIVES FOREVER

metro

Published by Metro Publishing
an imprint of John Blake Publishing Ltd
3 Bramber Court, 2 Bramber Road,
London W14 9PB, England

www.johnblakepublishing.co.uk

First published in paperback in 2009

ISBN: 978-1-84454-637-4

British Library Cataloguing-in-Publication Data:

A catalogue record for this book is available from the British Library.

Design by www.envydesign.co.uk

Printed and bound by Grafica Veneta S.p.A., Trebaseleghe (PD) - Italy

1 3 5 7 9 10 8 6 4 2

Papers used by John Blake Publishing are natural, recyclable products made from wood grown in sustainable forests. The manufacturing processes conform to the environmental regulations of the country of origin.

Images by kind permission of The National Portrait Gallery, p2, The Royal Society of Medicine, p3 and p7 and The Wellcome Library, London, p5, top and p7

Every attempt has been made to contact the relevant copyright-holders, but some were unobtainable. We would be grateful if the appropriate people could contact us.

Acknowledgements

When the first edition of this book, *Poison Arrows*, was published in 2005 it was warmly welcomed by the scientific press and those with an interest in medicine and physiology. It was the subject of BBC science programmes. The publisher, John Blake, insisted the story was too interesting and too important not to be accessible to a much wider audience of lay readers. And he persuaded me to write *From Poison Arrows to Prozac.*

In writing this book I would like to acknowledge the contribution of Stuart Robertson, of John Blake Publishing, whose suggestions and criticism have helped shape it; the assistance of my ever-critical proofreader, my wife Carole, and the patience of my friends, who have provided information, and my colleagues at Chelsea and Westminster Hospital, who served as guinea pigs for many of the experiments described in the book.

Contents

Preface

The words *toxins* and *toxicology* refer to poisons and the study of poisonous substances. These words are derived from the Greek *toxon*, a bow, and, when it is combined with *philos*, loving, it gives the sport of archery its name: *toxophily*. How, over the course of many years, bows and arrows came to be associated with noxious substances and poisons is the starting point of this book.

It is known that, since biblical times, hunters and warriors smeared their arrows with poisonous substances to make them more deadly. The practice is mentioned in the Bible and Homer refers to its use, in the *Odyssey*. With the discovery of gunpowder and shot the use of bows and arrows fell into disuse in Europe and Asia. However, guns and gunpowder did not reach the continent of America until well after Christopher Columbus discovered the New World and travellers and adventurers from Europe arrived with their weapons.

When these early voyagers to the New World returned home they told horrific tales of the lethal effects of the native darts and arrows. They told of how the natives smeared the arrows with a poison that they called '*ourari*'. It was these stories that caused bows and arrows to become so closely connected in people's minds with poisonous substances. Sir Walter Raleigh, in his *Discovery of the Large, Rich and Beautiful Empire of Guiana* (1596), told of the terrible agony suffered by those injured by a poisoned arrow; of the victim's 'staring eyes bulging out of their sockets with terror' and of 'bellies rendered asunder'.

They were convinced that the poison was produced by sorcery. According to their tales, it was made by a cabal of elderly women who they thought were witches. They believed it was the ritual associated with its production that gave the arrows a supernatural, magical killing power.

The story of curare

The story of curare has waited over three centuries to be told. It started with the reports about the arrow poison brought back to Europe by the explorers and adventurers who followed the Columbus trail to the New World. The belief that it had magic properties was fostered by the very secretive way in which the poison was made, involving an elaborate, semi religious ritual in the middle of the night.

They were both fascinated and terrified by the effect of the poison. They told of the mystical effects of this arrow poison called '*ourari*' and described the spell it cast over its victims. In his first-hand account of the effect of the poison, Sir Walter Raleigh, spoke of it as 'causing a death so horrible no man can abideth to see it'. This started the search for the secret of the poison and a means of frustrating its effect.

The discovery and taming of the arrow poison is a romantic story of exploration, self-interested curiosity, enlightened guesswork, scientific endeavour and serendipity. It is a story that spans three centuries.

It is the story of the experiments of the Abbé Felix Fontanna and Dr Brockelsby in Leiden in Holland, and of others in England. It tells of the demonstration by the famous surgeon Sir Benjamin Brodie, who showed that curare killed its victims by paralysing them and stopping them breathing. It was an experiment that did much to demystify the poison. It recounts the exploits of the eccentric English explorer and naturalist Charles Waterton, who brought large samples of the native poison to England for study. It tells of how, during his journeys in South America, he wrestled with a giant snake and rode on the back of an alligator so as to obtain specimens for his museum.

It is the story of the French physiologist Claude Bernard, who worked in Paris at the time of the Paris Commune. He is known as the father of endocrinology for describing the properties of insulin eighty years before it was actually isolated. He described cardiac catheterisation (inserting a catheter into the heart) almost a hundred years before it was used in medicine and talked of 'maintaining the stability of the internal environment' eighty years before the term *homeostasis* was introduced. He discovered how curare caused paralysis. His predictions have proved prescient. He used curare as a forensic tool to learn how the body controls the composition of the body fluids, which he foresaw as the key to life.

It tells how the English surgeon T. Spencer Wells and others used the poison to treat the convulsions of tetanus and rabies, both common diseases in Victorian times; and of how Richard Gill, an American living in Ecuador in the 1920s, saw the possibilities of using curare to relieve his painful muscle spasms, caused by a riding accident, much in the way we use Botox in pain clinics today.

Eventually, after many false starts, curare was introduced into medicine in 1942 as an adjuvant – in anaesthesia and psychiatry. Within a decade it had reduced anaesthetic mortality by almost one-third and was hailed as being as important an advance in anaesthesia as antisepsis had been in surgery. Today, it, or a synthetic copy, is used in the administration of the vast majority of anaesthetics.

Postscript

The story of curare does not end with its introduction into medicine and anaesthesia. There is an important postscript to the events described in the first part of this book, which have profoundly changed the practice of medicine and our understanding of the way the brain controls the workings of our body. It has resulted in the abandonment of the iron lung for the treatment of paralysed patients and an understanding of the importance of controlling artificial ventilation in order to bring relief to patients with many killing diseases.

It was during the process of finding out how the poison killed its victims, and the subsequent taming of the poison, so that it could be used in medicine, that important discoveries were made about the way our nervous system works. It settled, once and for all, a debate that had engaged many eminent scientists during the mid-twentieth century. The dispute centred on the various claims and counterclaims as to the manner in which the brain passes on its instructions to the body. It demonstrated the important role that chemicals, produced in the brain and nervous tissues, play in this process.

It would have been impossible for human development to have occurred without the evolution in this chemical messenger system. Without the flexibility provided by additional messenger systems the brain would not be able to meet the requirements of an increasingly sophisticated and complex lifestyle. It is through learning how curare affects these systems that the subtle changes that have occurred in these mechanisms during evolution have come to light. It is these changes that have allowed us to develop the range of responses necessary for our present lifestyle.

This new understanding has led to the development of drugs to control an excessive activity of these responses or to supplement a failing system, to perfect existing drugs and to develop new, more specific agents, with fewer side effects. It is probable that in the future it will lead to a cure for conditions such as the memory loss that occurs as we get older, Parkinson's disease, manic and depressive states, inappropriate emotional responses and many other disabling diseases.

This story that started with the South American arrow poison is not yet complete.

In telling this story in a way that is readily understandable I have tried to reduce technical jargon to a minimum, even though this has meant omitting details that might have been more satisfying to the tyro. I am aware that I have taken some leaps of faith. I have done so freely and in doing so I have promoted my own views about the way curare works. I have done so in a Popperian way,[1] aware that future research may prove them to be wrong. For these transgressions, I plead *mea culpa*; my excuse is that they provide a better explanation of what happens than any alternative, more conventional theory.

If I have entertained the reader with a new insight into an amazing, fascinating story, I will be satisfied. If I have occasioned him or her to look in a new light at the way the physiological processes of the body are controlled, I will have truly fulfilled the purpose of this book.

1 Karl Popper (1902–94) was an Austrian-born British philosopher who argued that scientific theories can never be proved to be true, but are tested by attempts to falsify them.

Prologue

Every one of the myriad cells that make up our bodies contains a minute drop of seawater. Not the seawater of today but as it was at the very beginning of life, in the vicinity of the underwater thermal vents that spew minerals into the ocean from below the Earth's crust. It was here that the first forms of life are believed to have evolved, thousands of millions years ago. The contents of this fluid, inside the cells, are the same whether it is from the cells, of the heart, the pancreas, the brain or the spleen. It contains a similar concentration of salts, such as sodium, potassium and calcium, and the same minute amount of oxygen. The nature of this salty fluid, the 'intracellular water', has to be maintained within strict limits. If the contents change significantly, all the various processes necessary for life will fail.

Not only is the formula the same for the fluid inside all the cells of one's body, but it is much the same as that in the cells of every mammal, irrespective of its diet or the climatic conditions in which it lives. We have much the same intracellular water as the elephant, the rat and the hippopotamus. This is no coincidence; it is the result of a common evolutionary history. It is a story that starts at the beginning of life on Earth.

Although no one can be certain as to the exact origin of life all the evidence points to it starting in the seas that covered much of Earth about 3,500 million years ago, on the ocean floor, where it existed as an immobile slime. It has been suggested that it originated with the formation of simple chains of chemicals, including a form of a primitive nucleic acid,[2] the harbinger of DNA, that crystallised out from the salts dissolved in the sea. Chemicals dissolved in water will form crystals spontaneously if the concentration and temperature is right, just as substances, such as sugar and salt, crystallise out of water at a given temperature and concentration. In order for this process to become the precursor of life it would have to have been continuous and the crystals themselves self-replicating.

What is generally accepted is that, after many millions of years, something happened that suddenly caused an evolutionary explosion about 600 million years ago. This was towards the end of the greatest ice age that the planet

2 It has been suggested that primitive peptide nucleic acid reached earth on meteorites and from the large amounts of interstellar dust that rained down upon its surface after it was formed. Meteorites and interstellar dust contain a high concentration of organic chemicals, including amino acids and purines, the precursors of DNA.

has ever experienced, when glaciers extended almost to the equator, the time of the so-called 'snowball earth'. It was at this time that many forms of multicellular life appeared.

What caused this extraordinary spurt in the development of life forms? Many explanations have been offered such as a change in temperature or in the composition of the atmosphere. The high CO_2 levels that had prevailed at the time of the great freeze (some 100 times the present concentration) declined and a significant amount of oxygen appeared, for the first time, in the atmosphere. However, the most likely explanation is that it resulted from the development of the cell membrane.

By trapping a tiny drop of water within a membranous, semipermeable envelope, the cell wall, it would have been far easier for the cell to maintain a stable concentration of chemicals within its internal environment. In this way it would have been able to provide the conditions for the continuous production of the chains of chemical substances that are essential for life.

Had it not been for another remarkable event, evolution might have become stuck at this point. Single cells, or thin sheets of identical cells, could control the contents of their internal water as long as they remained immersed in the constant environment provided by the sea. Any cells that were not directly bathed by seawater would have been in danger of drying out. What made the next step in the evolution of more complex forms of life possible was the appearance of special organs to monitor changes, such as drying out, and to link these to some mechanism that would allow it to protect itself. This required a communication system to be developed.

The various steps in this evolutionary pattern are speculative, but one can make a reasonable guess as to how they came about. Some indication is afforded by studying embryonic development and the stages that animals, such as the frog, go though in their maturation process. It starts with a single fertilised cell, the ovum, and develops into an embryo containing many apparently identical cells into a stage at which cells with differing functions start to appear. This allows its maturation into a fishlike tadpole, which lives in water and eventually emerges from its watery environment as a young, air-breathing frog.

It is probable that the only form of life that was capable of leaving the primordial swamp and surviving on land was composed of one or more simple, identical cells. This form of life would have survived by absorbing water and dissolved nutriment through the thin membrane covering the cell. The main danger to its survival would have come from changes in temperature or from a drying out of the environment. This dehydrating effect would have produced changes in the local concentrations of chemicals, such as sodium, potassium, chlorine and calcium, within the cell. It is believed that it was these alterations in the local concentration of salts that initiated the process of 'differentiation', in which new types of cells were formed. It was this process that led to the development of more complex organisms.

With the passage of many millions of years the

atmosphere became drier and temperature fluctuations more dramatic. Only those organisms that produced some means of adapting to these changes survived. Those that survived developed groups of cells that had special functions. These were the forerunners of the sensory and motor organs found in more advanced forms of life. Because of the drier environment in which they now existed, the surface of these early animals became covered with a thick skin or carapace to limit fluid loss. This impermeable covering would have prevented the organism from taking in fluid and nutriment through its surface and necessitated the development of special mouth areas through which food and water could be ingested.

As the environment dried out further and changes in the temperature become more extreme, the chances of survival depended on developing some means of locomotion to allow it to escape from a dry, hot area for a nearby site where water, nutriment and shade were available. This required some means of detecting changes in the environment, a sensory system, and a means of conveying the information from these sensory organs to the organs responsible for propulsion. For this to be possible a messenger system had to be developed.

Chemicals, acting as messengers, served this function. The first such chemical messenger must have been formed from substances that were abundant in Earth's atmosphere at that time, such as ammonia, carbon dioxide and water. It is probable that this was either acetylcholine or a similar substance. It would have been released in response to an alarm signal from cells capable of sensing changes in the outside world.

It is quite possible that the present-day role of acetylcholine, the most ubiquitous of all the chemical transmitters in the animal kingdom, dates back to these primitive times. It would have functioned in much the same way as it does today in the clams and oysters on our seashores. In these bivalve molluscs, acetylcholine is released into the fluid bathing the tissues in response to a signal produced by touching the shell. It causes the strong muscle that closes the two halves of the shell to contract.

Even today an atavistic enzyme that destroys acetylcholine is present in the bloodstream of man. No physiological function can be ascribed to this enzyme; indeed, a number of people live perfectly normal lives without it. It is possible that its presence is a footprint of the evolutionary stage when acetylcholine was released directly into the tissue fluids.

As evolution progressed and these organisms became more complex, special structures capable of detecting changes in the outside world – such as eyes and ears – developed in parts of the body that were remote from the muscles involved in making an appropriate response. This necessitated a more sophisticated means of communicating information from one part of the body to another. This need was met by the development of cord like nerve trunks as simple extensions of nerve cells. They connect the sensory organs to the brain and carry instructions from the brain to other parts of the body.

By this means it was possible for the transmitter, acetylcholine, to be released at nerve endings some distance from the sensory organ itself but close to the sites responsible for initiating a response. Acetylcholine then became a specific transmitter, passing on the messages carried in the nerves to the muscles, heart and circulation.

The next stage in this story occurred when animal life evolved still further and started foraging for food and shelter. Newer transmitter substances appeared that helped prepare animals for this new challenge. Later still in the evolutionary development of the brain, we find more sophisticated transmitters appearing associated with the response to hunger, mood, sexual arousal and memory. However, acetylcholine continues to provide the background activity that controls the release of all of these substances.

Curare

Without chemical messengers, the evolution of complex forms of life would have been impossible. The first intimation that such a system might exist came from the French physiologist Claude Bernard in the nineteenth century. He suggested that the ultimate purpose of all the control systems of the body – which are constantly in action, although we are not aware of them – was to maintain the integrity of the bodily fluids within the narrow limits necessary for life. This he proposed was the essential condition for *la vie libre* – life in a constantly changing external environment.

This was the position until about sixty years ago. By 1930, evidence was emerging that the chemical acetylcholine was involved in adjusting the heart rate and that it was also released when muscles responded to nervous commands. Over the next two decades there was an acrimonious debate as to the role it played in these responses imitiating them, while others ridiculed the idea.

It was at this point in the argument that the investigation into the role of acetylcholine in muscle activity converged with studies into the way in which the South American arrow poison, curare, killed its victims. There was a continuous thread in the experiments involving curare which stretched back over four centuries. It started with the early explorers to the New World, who described the terrible manner of the death of those poisoned by darts anointed with this poison and continued intermittently until the 1930s, when its action was finally demonstrated.

This was one of those coincidences that occur from time to time in scientific investigation when a discovery in one discipline provides the means of making a quantum leap in another. In this case it was the understanding of the way curare paralysed its victims that allowed the researchers at University College in London to demonstrate, unequivocally, that the brain controlled the functions of the body by means of chemical messengers such as acetylcholine. It was a discovery that changed the face of medicine.

Prologue

This story begins with the discovery of America by Christopher Columbus in 1492 and the subsequent voyages of explorers to the jungles of South America.

CHAPTER ONE

The South American Arrow Poison

A spice is a dried seed, root, bark or fruit used as a food additive for flavouring and indirectly for preventing putrefaction and the growth of pathogenic bacteria.

Spices, such as cloves, have been used to flavour foods since ancient times. The Bible tells, in Genesis, how Joseph was sold into 'the slavery of spice merchants' by his brothers. There are frequent references to the use of clove oil in Roman literature, where it was used to mask body odours and for religious rituals. However, until the Middle Ages, the practice of using spices in the preparation of food was largely restricted to the Middle East. In these countries, cinnamon, black pepper, cloves, saffron, cassia and ginger were used by those wealthy enough to afford them.

It is probable that the taste for these spices was brought to Europe by the Crusaders. At first they were used mainly by the rich merchants of France and Spain, but eventually their use spread to England. By the fifteenth century the demand for these spices had spread throughout the whole of northern Europe, although they were always too expensive to be used in everyday cooking. In spite of their cost, the demand for pepper, cinnamon, cloves, nutmeg and ginger was huge.

By the end of the fifteenth century it was estimated that about 1,000 tons of pepper and 1,000 tons of other spices were imported into the Port of London each year. Pepper was widely used to disguise the flavour of meat from animals that had been slaughtered before the onset of winter and kept from going bad by salting and pickling. Without the addition of a spice the meat was invariably so salty that it had to be soaked in water to make it eatable.

So it was that, in the fifteenth and sixteenth centuries, shops selling pepper and other spices and merces were commonplace in the streets and alleys of towns such as London. The trade in pepper was so important that it led to the establishment, by Royal Charter, of the Company of Pepperers. It eventually became a part of the newly formed Grocers Company.

Unfortunately, by the second half of the fifteenth century, pepper was in short supply and the price had escalated, putting it out of reach of all but the very wealthy. Part of the increase in cost was due to a growth in demand, as it became more and more fashionable, but most of the price rise was due to the insecurity and expense of its transportation to England from the Orient and the Far East, where it was produced.

Almost all the spices came from the Moluccan (or Maluku) Islands of the East Indies – known as the Spice Islands – and from the southern coastal areas of India and Serendip (Sri Lanka). At its source in these countries a bale of pepper would cost less than one guinea (£1.05) but, by the time it reached the shops in London, it would sell for more than a hundred times that amount. This was due to the long and hazardous journey necessary to bring the spices to Europe.

Originally, most of the spices travelled by the Silk Route across Asia to Constantinople (Istanbul), but this route, largely exploited by the traders from the Republic of Venice, became increasingly difficult after the fall of Constantinople to the Ottoman Caliphate. The alternative route started in the Orient, where bales of spices were loaded onto boats and carried by Arab traders across the Red Sea. They then travelled overland to the ports of the Mediterranean, such as Alexandria. A large part of this land route was controlled by Arab warlords, who either extracted a toll or required a duty to be paid when goods passed through their lands.

There was a particularly sharp rise in the price of all the spices following the fall of Egypt to the Ottoman Empire in 1453. This gave them control of the all the Mediterranean ports of the Levant and Egypt. The problem was made worse because at this time the Ottoman Empire was at war with Catholic Europe.

Christopher Columbus

By the middle of the fifteenth century only the very wealthy could afford to use spices. Their use became restricted to great banquets and grand occasions. Because of the increase in demand, there was considerable commercial interest in finding an alternative way of bringing these spices from the Orient to Europe. This caused an upsurge of interest in the possibility of finding a sea route from Europe to the Indies and the Far East.

It resulted in some of the great voyages of exploration that took place at this time. It led Henry, King of Portugal, to establish a school of navigation at Cape St Vincent, the most westerly point of the European continent, to train sailors and to encourage exploration of the coast of Africa in the hope of finding a way to the Indies and the Spice Islands.

In the early part of the fifteenth century, Portuguese and Spanish sailors sailed down the coast of Africa in the hope of finding a way around this seemingly endless land mass. They also voyaged across the Atlantic as far as Madeira and the Canary Islands, where they established colonies. In 1488, Bartholomew Diaz, a Portuguese explorer, returned to Lisbon having found the southernmost tip of the African continent. His rapturous reception was not only a tribute to his navigational skills but also recognition of the commercial implications of his voyage. It opened up the possibility of an alternative route to the Spice Islands by sailing round the Cape of Good Hope and to the East. It eventually led to the voyage of

Vasco de Gama in 1499 and the establishment of a Portuguese colony in Goa, India, in 1510.

It would not have been at all surprising if the Genoese sailor Christopher Columbus had been dockside in Lisbon when Bartholomew Diaz returned home in 1488. At the time of Diaz's voyage, Columbus had left his native country and settled with his wife and children in Lisbon, where, together with his brother Batholomew, he earned his living as a map maker. He no doubt listened carefully to the accounts of the voyages made by these early explorers and incorporated the information gleaned from their stories into his maps.

There is little doubt that most navigators at the time realised that the world was a sphere, but they had little idea of distances involved or the land masses that existed. Many believed that there was open sea between Europe and Asia, and sailing to the West from Europe would land them in Asia and the East Indies. This view was the result of the writings of Ptolemy, who considered half the surface of the world to be covered with water and the other to be composed of the land mass of Europe and Asia. (Fig 1)

Columbus became convinced, after studying the maps available at that time and listening to the tales told by sailors returning to Lisbon, that it would be possible to reach the East Indies by sailing some 2,400 miles to the West across the Ocean Sea, as the Atlantic was known. He came to this conclusion from calculations based on the erroneous notion prevalent at the time, that the sphere of the Earth was 25,255 kilometres in diameter. He concluded that the 2,400 miles to the Indies was 'not too great a space to be passed' and that such a voyage 'has become not only possible but certain, fraught with inestimable hazard and gain and most lofty among Christians'. The problem he faced was mainly one of logistics: how to carry sufficient water and food for such a long time at sea.

In his opinion, presented in his numerous petitions to the Spanish and Portuguese courts for royal patronage, he argued that this would be no further than sailing to the south around the tip of Africa and then to the East and across the Indian Ocean, and it was likely to prove less hazardous. However, his attempts to interest the Portuguese king in the adventure failed, since he could not persuade the cautious Portuguese that he could overcome the difficulty of carrying sufficient food and water for such a long time at sea.

After the failure of his attempts to obtain royal consent from the Portuguese – who by the latter part of the fifteenth century had established a vested interest in the passage around Africa to the Far East – he turned to Spain. After an initial failure, he finally persuaded King Ferdinand and Queen Isabella of Spain to back the venture. It is recorded that this was largely due to the intervention of a converted Jew, Luis de Santangel, who was the keeper of the queen's purse and who probably drew up the rigorous terms under which royal patronage for the voyage was provided. Certainly, by 1492, when approval was granted, the king needed a source of money as the war to expel the Moors from Grenada and southern Spain had exhausted the treasury. Under the contract that was drawn up, half the cost of the expedition had to borne by the citizens, while the crown provided the rest of the money and received the lion's share of any bounty.

As dusk fell on the evening of 3 August 1492, Columbus's fleet of three small ships – the largest one, the *Santa Maria*, commanded by the newly promoted Admiral of the Oceans, Columbus, and the two caravels, the *Pinta* and the *Niña* – set sail from the port of Palos in southern Spain. Their first stop was the Canary Islands, where they obtained fresh food and water before setting out to sail across the ocean.

After five weeks at sea, they eventually sited land at 2 a.m. on 12 October 1492 and anchored off what is probably one of the islands in the Bahamas. They were fortunate in the choice of the time of year they chose to cross the Atlantic. Even today, yachtsmen sailing to the West Indies try to leave from the Canary Islands towards the end of September so as to pick up the favourable westerly winds prevalent at that time of year.

Clearly, it was not purely altruism or even the desire to establish new overseas colonies that persuaded the Spanish crown to support the expedition. The Spanish monarchs needed money to pay for the war against the Moors and the looming wars in Europe. Although this was a high-risk adventure, it promised enormous rewards if a secure route to the Spice Islands could be found by sailing across the Atlantic and spices could be brought to Europe through the ports of Spain.

When the three small caravels made landfall in the Caribbean, Columbus was confident that he had arrived in the East Indies. Even by the time he returned to Spain, he was unaware that he had not landed in the East Indies.

During the time he spent in the Caribbean he made several exploratory trips. On one such voyage he discovered Cuba (Hispaniola) and established a colony there. On his return to Spain, he painted a glowing picture of the fertile and beautiful land he had discovered, whose rivers contained gold and whose mountains were a rich source of minerals. He also suggested that he had evidence of countries that were rich in silver and gold that were a short distance to the west of the land he had found.

The Columbus trail

A combination of missionary zeal and the lure of gold led to further expeditions in the years following Columbus's return from his first voyage to the New World. Very soon Spanish colonies were established on the island of Hispaniola and at Darien, on the southern end of the Panama isthmus. It was from this latter colony that Hernán Cortés led his expedition to the land of Montezuma and the Aztecs, in what is now Mexico. He sent back gold and silver treasure to King Charles in Spain as evidence of the riches to be found in the New World.

However, it was the exploits of Francisco Pizarro, a

noble of Salamanca who became an influential figure in the disputatious court of the governor of Hispaniola, that fired the imagination of the Spanish court. Although Amerigo Vespucci and Francisco Pizarro had explored the coast of South America as far as the origin of the Amazon, they had met hostile savages who had denied them access to water and victuals necessary for a prolonged exploration.

By a feat of great seamanship, Pizarro travelled much further down the South American coast and, after many false starts, made his way inland to discover the secret of the Incas in the heights of the Peruvian Andes. The enormous quantities of gold and silver sent back to Spain from his conquest led to credible stories reaching Europe of cities of gold and silver and of untold treasure to be had by those bold enough to venture to this New World.

The arrow poison

When the news of the gold and silver being brought to Spain reached the sea ports of Europe, it caused feverish excitement. Many ill-prepared adventurers soon set out on the perilous voyage to the New World. For the first time in Europe money was available to back these schemes. Sir Thomas Gresham had built his Royal Exchange near the stock market in London and risk companies were established to fund 'merchant adventurers of good report'. Although most of the voyagers who set out for the New World were privateers and buccaneers seeking their fortunes, there were also a number of soldiers and monks among them eager to convert the heathen and to establish dominion over this new land.

Inexperienced adventurers, many of whom had never been beyond their home town or village, crowded into small, poorly constructed vessels that were unsuitable for the long ocean crossing. They knew nothing of the prevailing winds and storms in the southern ocean, the extremes of climate they would have to endure or the privations they would suffer. Many died of dehydration, scurvy, disease and starvation on the long sea journey. The vessels were seldom big enough to carry sufficient provisions.

Many never reached the New World. Of those that did a large number suffered scurvy, typhus and swamp fever. One can only imagine the fear and wonderment with which those who survived the journey viewed this foreign land. As they landed, they found themselves surrounded by hostile natives, armed with spears and blow darts, who would suddenly emerge from the seemingly impenetrable jungle lining the shores, with their faces and bodies covered with paint and feathers. (Fig 2) They found ferocious animals and brightly coloured birds that were unlike anything they had ever seen before. They had to cross muddy rivers swarming with flesh-devouring fish and with alligators that bit off the limbs of the unwary. They encountered poisonous snakes and swarms of stinging insects whose bites produced sores and fever.

They brought back colourful tales of their adventures to amaze and impress their sponsors. None of these tales was more astonishing, or caused more concern, than that of the mystical properties of the substance into which the natives dipped their darts and arrows. They described, in some detail, 'the flying death' caused by the magical South American arrow poison. (figure 3)

The practice of anointing arrows and darts with poisons was not new: it had been common practice in ancient times. Homer refers to their use, and in Virgil's *Aeneid*, Ovid is described as smearing his arrows with 'viper's blood' and Apollo's darts were said to carry 'pestilence'. The word *toxin* is itself derived from the etymological stem *toxon*, meaning a bow or a bow and arrow. There is evidence that the practice of dipping arrows into concoctions of poisonous herbs was common in Europe among the Celts and Gauls, but that it had died out with the advent of more effective weapons and the discovery of gunpowder. From the stories told by the explorers of the New World, the effects of the poisoned arrows were at first thought to be some form of witchcraft. It was only when the natives were observed dipping their darts into the sticky concoction of poisonous herbs that the reason for their lethal effect became obvious.

This lethal effect was consistently reported by voyagers who returned from these jungles. Especially terrifying tales were told by those who ventured along the Amazon River and into the Orinoco Basin, where the bloodthirsty warriors, native to this region, used a particularly potent form of the poison for hunting their prey and killing their enemies.

It is the tales from these explorers that contained the most graphic details of the 'flying death' wreaked by the blow darts used by these Indians. Many of the published reports were written to titillate a European audience hungry for excitement, who longed for extraordinary stories of exotic discoveries and flesh-creeping tales of the dangers of this New World. It is not surprising that some of the stories were purposely embellished and made absurdly fanciful to increase their appeal to an expectant and gullible audience and to boost the sales of their books. As a result it is difficult to distinguish where fact ends and mysticism begins.

Clearly, the death caused by the poison was horrendous. Stories of the victim dying with 'staring eyes', of his 'bowels exploding', of his being in 'convulsions' and 'fixed to the ground unable to move' are so common in published reports that they must, in large part, be believed. One of the most frightening aspects of the nature of the death was that there appeared to be nothing that could be done to remedy the effect of the poison. It soon became apparent that there was a pressing need for an antidote with which to treat the victims, but first more had to be known about what the poison was and how it killed its victims.

Possibly the earliest details of how it was produced came from the account of the Italian monk, Peter Martyr d'Anghiera, who joined an expedition established to convert the heathen natives to the Catholic faith. He published an account of his report to the Spanish court early in the sixteenth century. The book, *De Orbe Novo* ('Of the New World'), which was translated by Dr Mac Nutt of New York, tells of a poisonous concoction being made by members of a small cabal of female elders who were kept sealed in a special hut for one or two days. He wrote that 'they often died from inhaling the fumes.'(Fig 4)

If all the women survived the ordeal it was considered that the poison was not sufficiently strong. According to a native high priest, it was made from 'the stings of scorpions, the heads of deadly ants and a juice distilled from special trees'. His account of the death of those making the poison is a recurring theme in reports from other sources. The stories of these deaths may well reflect the extreme precautions taken by the high priests to guard the secret source of the poison and the process by which it was made. In all probability this involved deliberately killing those who became privy to the secret, as an offering to the gods, since the practice of human sacrifice is known to have been common in this region at this time.

There is also a suggestion in *De Orbe Novo* that the natives had developed a cure for the effects of the poison, which, according to Peter d'Anghiera, involved avoidance of 'intoxicating drinks and substances and of the excessive pleasures of the table' and called for 'sexual abstinence for two years'.

In the years following Peter d'Anghiera's account of his voyage to the New World, other explorers ventured up the massive Amazon to try to reach its source. They brought back accounts of their exploits together with curios and samples collected from the tribes of Indians who inhabited the forests lining the riverbanks. One of the earliest of these adventurers was the Spanish physician, Nicolás Bautista Monardes, who wrote an account of his adventures in 'purple prose' (but in Spanish) in his book, *Dos Libros*, published originally in 1565. The book was translated into English by John Frampton in 1571, under the seemingly unsuitable title of *Joyfull Newes from the Newe Founde Worlde*.

In his book, Monardes also tells of the dangers of inhaling the fumes of the arrow poison and of the horrible death it causes. He described the mysterious ritual that surrounded the preparation of the poison and, he believed, gave it magic powers. Like d'Anghiera, he probably mistook the extreme precautions that were taken to preserve the secrets involved in the preparation of the poison for a magic rite.

More accurate and objective accounts of the mysterious arrow poison and the manner in which it killed its victims followed the return of Sir Walter Raleigh's expedition to the Guinea lands (Guyana) in South America. In 1595, Sir Walter Raleigh wrote:

> I was curious to find out the true remedies of the poisoned arrows... the person shot endureth the most insufferable torment in the world and abideth the most insufferable death... No man can endure to

> cure them... there never was a Spaniard either by gift or torment that could attain the cure even although they have martyred and invented torture.

One member of the expedition, the famous explorer Dr Richard Hakluyt, made a second voyage to Guinea, after which he described the preparation of the poison by a special priesthood who supervised the collection of the ingredients and officiated in the rites associated with its preparation. He also believed that the natives had an antidote for the poison, but neither he nor anyone else was able to determine the secret of their cure.

It is apparent from his account that the Spaniards frequently used torture to try to learn their remedy. His description of the suffering of those dying from the effects of the poison makes it clear that the victims remained conscious with staring eyes but were unable to cry out during the prolonged agony of their death. His horror at the terrible torment they suffered is all the more surprising when one considers that death by disembowelment, hardly a pleasant manner of dying, was practised as a punishment in England at that time.

Such was the fear created by these stories, it was evident that there was a pressing need to learn more about the nature of the poison and why it produced such a terrible death. The first step was to obtain samples of the poison that could be brought back to Europe, where it could be studied in a scientific atmosphere. Only in this way could its magical properties be assessed.

CHAPTER TWO

Unravelling the Mystery

The earliest recorded experiments on specimens of the arrow poison, 'ourali' or 'wourari', were carried out in Leiden, Holland.

The University of Leiden was formed in 1575, after William the Silent, Prince of Orange, raised the siege of the city. In return for the steadfastness of its citizens, he offered them a year free of taxes or a university. They chose a university. The university soon became an important centre for scientific learning and experimentation. The Low Countries were essentially a maritime kingdom at this time with a large proportion of the population engaged in seafaring trades. Indeed, so great was its fame as a centre of seamanship that Jonathan Swift, in *Gulliver's Travels*, had Gulliver sent to Leiden to study navigation before embarking on his adventures.

It would have been surprising if a major maritime nation, with a history of successful voyages to the East

Indies, had not participated in the exploration of South America. It is probable that the first samples of the crude arrow poison were brought to the Low Countries by a voyager from one of these expeditions. It is recorded that it was being used in experiments in the University of Leiden as early as 1740.

The first recorded experiments were those of De la Condamine (1701–74), who, in the early part of the eighteenth century, gave the poison to pullets in the hope of finding a remedy for their lethal effects. Stories brought back by voyagers, including Hakluyt, had led him to believe that sugar and salt were antidotes to the arrow poison. His experiments were hardly convincing, but he recommended salt in preference to sugar! We now know that both salt and sugar are useless against the effects of curare.

Another investigator, working with the same parcel of poison in Leiden at about the same time, was an Englishman, Dr Brockelsby (1722–97). He showed that, even after the poison had been injected into the leg of a cat and 'its breathing had appeared to cease', its heart continued to beat for almost two hours before it finally died. This experiment demonstrated for the first time that the poison did not kill by stopping the heart.

By 1745, when the Italian Abbot, Felix Abbada Fontana (also known as Felice Fontana, 1720–1805), performed his experiments, Leiden was the most famous centre of scientific enlightenment in Europe. Fontana was a distinguished anatomist who became the director of the

Natural History Museum in Florence. He is remembered today for his description of the spaces of Fontana in the eye. He had been attracted to Leiden by its reputation for intellectual freedom and scientific experimentation.

His experiments, carried out on chickens, showed that the fumes of curare did not kill or injure the birds and that their flesh was not tainted or rendered unfit to eat. However, merely piercing the skin with a lance tipped with arrow poison rapidly killed them. These experiments clearly demonstrated that the fumes of curare were not lethal and that the stories spread by the early explorers of deaths from inhaling the fumes were wrong. They were, in all probability, a cover for the ritual killing of those involved in the preparation of the poison. There is little doubt that this was carried out in order to maintain the secrecy that surrounded the preparation of the poison.

Charles Watertom (Fig 5)

It was the English naturalist and adventurer, Squire Charles Waterton (1782–1865) who first described the poison as 'wourali', the name given to it by the natives in the Esquibito river region of South America. In his book *Wanderings in South America* he described his adventures and observations during his journeys in the Amazon jungle in the company of his 'Indian' Machousi guides and a freed African slave called Dadi. Although he was principally interested in the great variety of bird life he found there he made a detailed study of many animals, including sloths, monkeys, caymans (which are like alligators) and snakes. It was during the first of his four trips to South America that he observed the Indians using arrows tipped in poison to catch their prey. It was as a result of this interest that he came to play an important role in the curare story.

If you were to visit the town of Wakefield in West Yorkshire in England you might come across Waterton Road and Waterton School, both named in honour of the remarkable, eccentric squire of nearby Walton Hall. Charles Waterton was born at Walton Hall, into an aristocratic, Roman Catholic dynasty with links to several European royal families who claimed they could trace their origins back to Edward the Confessor.

One of his ancestors, John de Waterton, served as Master of the King's Horse at Agincourt. Unlike many aristocratic families, his forebears refused to convert to Protestantism during the reign of Henry VIII.

Charles Waterton was educated at the Catholic Stoneyhurst College and remained a devout, ascetic Catholic throughout his life. He lived an active life to a ripe old age in spite of repeated attacks of 'ague' (probably malaria). Even when he was over eighty, a neighbour saw him climbing a tree to return a bird that had fallen from its nest. He died following a fall on his estate, in which he fractured several ribs and damaged his liver.

In 1812, as a young man of thirty, he made his first voyage to Guyana, where his uncle had coffee plantations. He set about establishing one of the finest collections of preserved birds and animals in Europe, using his own preserving technique to maintain the colour and structure of the animals. He maintained that stuffing them distorted their shape. Many of the preserved specimens can be seen today in the museum at Wakefield. They are still colourful and lifelike.

It was the demonstration by the natives of the lethal effects of the dart and arrow poison that caused Charles Waterton to become interested in 'wourali'. He described his travels, his discoveries and his vicissitudes in Guyana in his highly successful book *Wanderings in South America*, first published in 1879. It is written in the manner of a diary and, in spite of some exaggerations and embellishments, it very soon became a bestseller.

There is little doubt that Charles was an odd man. He was tall and thin and of a somewhat domineering manner. He was an astute observer and a resourceful inventor, but he was also very eccentric. He married Anne Edmonstone, who was descended from Arawak royalty. He tells how he fell in love with her when she was a baby and waited until she was seventeen before marrying her. When she died in childbirth he was so grief-stricken that he vowed never to sleep comfortably ever again. From that time on he slept with a wooden block as a pillow.

Waterton is credited with having introduced bird boxes into Britain. At first, this was to try to nurture little owls, but later they became widely adopted on the estate.

Reading his story leaves one with the impression that he was a restless adventurer who, like many wealthy men of his era, became interested in natural science. This interest was heightened by his experiences on his journeys in Guyana and his expeditions in both South and North America (he was particularly taken with the civility and elegance of the Americans and with the handsome buildings that lined Broadway in New York).

He describes how, with the help of the slave Dadi, he set about collecting animals and birds, with the minimum of damage, so that they could be dissected and preserved. It was due to his skill at taxidermy, using mercuric chloride to harden and preserve the skin, that he was able to establish a huge collection of preserved (he insists none of the specimens were stuffed) exotic animals and birds, including monkeys, sloths, parrots and alligators for his menagerie at Walton Hall.

His enthusiasm as a collector knew no bounds. He describes how he was determined to capture a large cayman and how, after several attempts to snare the animal using a large hook baited with meat, he eventually came up with a special device consisting of a pole armed with many hooks, like a cat-o'-nine-tails. When, after many nights of anticipation, the animal took the baited hook and was impaled, he leaped onto its back as it was pulled to the riverbank.

The animal was duly dissected and preserved and can be seen to this day in the museum at Wakefield. The account given by one of the Indians who assisted in this capture suggests that it had its head in a snare and was half dead when this particular episode occurred. On another occasion Waterton wrote of how he fought with a huge snake, which all but squeezed the life out of him, in order to overcome it and add it to his collection. He describes his disappointment at failing to have himself bitten by a 'vampire' (blood-sucking bat) in spite of sleeping exposed for several nights.

During his various journeys he suffered from repeated bouts of malaria and had few qualms about bleeding himself to relieve the ensuing fever. He sustained several quite serious injuries during his adventures and endured extremes of temperature and humidity. Unlike many of his contemporaries, Waterton travelled without personal servants, relying on assistance from the natives. He carried no tent, only a sheet under which he sheltered when it rained and when he needed to carry out his dissections of the animals he had captured.

There is no doubt that Charles Waterton was a strong, brave man and a fearless adventurer, but he was not above using exaggeration in his stories so as to make them more exciting for the reader. This had the unfortunate effect of making the scientific community suspicious of many of his observations and confirming his reputation as an oddball. This reputation was not helped by the reports from visitors to Walton Hall that, on occasions, Waterton would sit under the table growling and pretending to be a dog. On one occasion, while in his dog persona, he is said to have bitten the leg of a visitor. It is reported that he tried to fly and launched himself off the roof of an outhouse, proclaiming he was 'navigating the atmosphere'; while his physician Dr Hobson tells of coming across the seventy-seven-year-old Waterton scratching the back of his head with the big toe of his right foot in the manner of a monkey.

Because of these oddities it is easy to dismiss Waterton's achievements, but his contributions to science are real and show evidence of a talent for detailed scientific observation.

In his studies on the behaviour of birds he records 119 species that he found in the grounds of Walton Hall. His efforts at animal and bird conservation and his descriptions of the feeding, mating and behaviour of birds, and the exclusively arboreal habitat of the sloth, have stood the test of time. The description he gave in his reports of the different species of monkey demonstrate his patience and objectivity as an observer of animal life.

He built a 9-foot-high wall around the 3-mile perimeter of Walton Hall (he said he paid for it from the wine he did not drink!) in order to allow him to establish, what was probably the world's first nature reserve, on his estate. He fought the first legal battle in England over an environmental issue. He accused his neighbour, a soap manufacturer, Mr Simpson, of polluting his lake. Unfortunately, he died before he had the satisfaction of having the nuisance removed to another part of Wakefield, by court order. However, he will always be remembered for his interest in wourali (curare), for without him and the specimens of the poison he brought back to England this story might not have progressed as it did.

Waterton made four separate voyages to the New World. In *Wanderings in South America* he gives an account of his journeys. It contains a detailed description of the effects of the South American arrow poison. He described seeing a wild pig killed by the natives using poisoned arrows. He tells of the manner of the pig's death in some detail: 'It affects the nervous system and thus destroys the vital functions of the body… the pig managed less than 200 paces before it dropped dead.'

After his second journey, he brought a large sample of wourali back with him to England. He used some of this specimen in an experiment on an ass. He injected a dose of the poison into a leg of the animal after tying a ligature around the leg a little above the site of the injection. He observed that 'the animal walked about as usual, and ate his food as though he were well. After an hour had elapsed, the bandage was untied and ten minutes later death overtook him.'

Waterton is not forgotten, and the Waterton Society helps to keep alive the legacy of this truly remarkable eccentric

Benjamin Brodie (Fig 6)

Waterton enjoyed showing visitors to Walton Hall his collection of wild and stuffed animals. Because of his interest in natural science, he mixed widely in the scientific community. He was a close friend of Professor Sewell, who was the president of the Veterinary College; they were both members of the Royal Society. It is probable that it was through his friendship with Sewell that he met Sir Benjamin Brodie (1783–1862), a distinguished surgeon, who was also a member of the society.

Brodie was a close friend of the eminent naturalist Sir Joseph Banks, who had accompanied Charles Darwin on the voyages of the *Beagle* and who became president of the Royal Society. It would have been natural for them to have shared an interest in Waterton's accounts of his adventures in the jungles of South America and of his observations on the lethal effect of the arrow poison. About this time Brodie had been studying the effects on the body of various poisons and later went on to publish his observations in a well received paper on 'The Effects of Certain Vegetable Poisons'. It is generally assumed that Waterton supplied Brodie with samples of the wourali poison that he used in his experiments when he returned to England after his first journey to the New World in 1812; but Brodie makes no reference to this in his dissertations and gives credit to a Dr Bancroft as being the source of the poison.

Brodie was altogether a different sort of person from Waterton. Whereas Waterton was flamboyant and adventurous, Brodie was conservative and cautious. He was a deeply ambitious man who from an early stage set his eyes on achieving great office. He was related to the Lord Chief Justice, who played a prominent role in the prosecution of Queen Caroline.

Unlike the dilettante Waterton, Brodie was said to have been 'consumed with a rage for work'. He was made a fellow of the Royal Society in 1810 while an assistant surgeon to St George's Hospital in London. He was appointed surgeon to the hospital in 1822. He attended

King George IV and became his medical confidant, spending hours, at a time, at his bedside. He became president of the Royal Society and the Medical and Chirurgical Society of London (now, after a royal charter and mergers, the Royal Society of Medicine) and was the first president of the newly formed General Medical Council. He was made president of the College of Surgeons in 1844 and contributed many specimens to its anatomical museum.

Brodie was an experienced anatomist and surgeon. His most famous work was on injuries and infection of bony joints. This led to a more conservative approach being adopted in the treatment of bony fractures and infections. He is remembered today for the eponymous Brodie's abscess, a painful swelling of a bone caused by a chronic infection. His papers tell of his investigations into the influence of the brain on the action of the heart, the actions of poisons as well his studies on bones and fractures.

Some two years after he was elected a fellow of the Royal Society, he enthralled its members with his now famous demonstration of the action of the vegetable poison wourali. It is typical of the man that, before performing in front of this illustrious audience, he privately carried out similar experiments on small animals to make sure the demonstration would be successful. He described in two papers, given to the Medical and Chirurgical Society of London, experiments he carried out with wourali on guinea pigs, cats and rabbits. Once convinced he could reproduce its effects, he arranged for the demonstration before the members of the Royal Society in London in 1812.

In this experiment it is reported that he brought a 'she ass' into the lecture theatre. He showed that the arrow poison, which on this occasion he termed 'woorari', killed its victim. It is recorded in the minutes of the society that, shortly after the injection of a small dose of woorari into the leg of an ass, it caused the animal to become paralysed and to stop breathing.

He performed an immediate tracheostomy, an operation with which every surgeon was well practised in the days of diphtheria, and inserted a domestic bellows into the animal's windpipe. This allowed him to maintain the animal's life by inflating its lungs with air (in this he was 140 years ahead of his time, for it was only during the polio epidemic of the 1950s that positive-pressure artificial ventilation was used to overcome respiratory paralysis). The minutes of the society state that 'he restored the life to a she ass poisoned by woorari by rhythmically pumping air into the animal's lungs for two hours'. It is also recorded that the animal survived the experience and lived content for several years after the experiment.

It was by means of this demonstration that Brodie showed how curare killed its victim. Unlike other more common poisons, it did not kill by poisoning the brain or the heart but by causing paralysis of the muscles of respiration, resulting in death from asphyxia.

Although there is good evidence that Charles Waterton

also performed a similar experiment on an ass he called 'Wouralia' with a similarly successful outcome at about the same time, he failed to convince the scientific community of the importance of his observation. This may have been due to the lack of public awareness of his experiment, but is more likely to have been due to scientific scepticism concerning his work. Although there is some uncertainty as to whether Brodie or Waterton was the first to demonstrate that artificial ventilation was capable of maintaining life in animals poisoned with curare, Brodie's presentation at the end of 1812 was undoubtedly more scientifically significant than the anecdotal account of Waterton's experiment.

What is the arrow poison?

One of the problems that dogged the early experiments with the South American arrow poison was the variation that occurred in the amounts of the different ingredients that went into the samples used and the question of whether it deteriorated with age. Benjamin Brodie recognised this and commented that his woorari was similar to that supplied to the Abbé Fontana for his experiments.

The poison certainly varied according to the region in which it was produced. Specimens collected from the Orinoco basin by the Schomburgks were largely composed of the bark and root of the creeper *Strychnos nux*. (Fig 7) Indeed, they called the poison *Strychnos toxifera*. As a result of its strychnine content, the immediate convulsive effects predominated, concealing the onset of the paralysis produced by the curare. This was probably why the German physiologist Albert von Bezold, using a preparation given to him by the Schombergs, believed the drug killed its victims by inducing convulsions. The poisons prepared from the more western jungles of Ecuador and the Peru were principally composed of an extract of the bark of vicuñas, and particularly that of the creeper, *Chondrodendron tomentosum*. (Fig 8) These contained high concentrations of curare and were similar to those collected by Waterton and those used by Brodie to demonstrate the paralytic properties possessed by their toxins. In an attempt to reduce the confusion, Dr Rudolph Boehm, in 1886, suggested classifying the curares according to their mode of preparation in the belief that the means used to store the poison indicated the region from which it came. He separated the curares into pot curare, tube curare, and calabash or gourd curare. This classification reduced the confusion, but it was not until 1935 – when Harold King, working in the Burroughs Wellcome laboratories in England, isolated and identified the chemical structure of curare – that pure specimens of the drug became available for study.

By the middle of the nineteenth century supplies of crude tube and calabash curare, of a reasonably consistent quality, were available in Europe thanks to the work of Wilhelm Peyer, who prepared a crystalline salt of the poison. Although this was later shown to contain potassium carbonate and phosphate as impurities, it was much better than the crude preparations previously available.

Unravelling the Mystery

By this time it was appreciated that curare caused death by paralysing the muscles of ventilation, and that, provided artificial ventilation was maintained, it did not cause any long-lasting ill effect on the body. The riddle of how and why it produced paralysis was to wait until the simple, elegant experiments of Claude Bernard in France, the intuition of the physiologist Otto Loewi and the experiments of the Henry (later Sir Henry) Dale and his team of fellow pharmacologists at University College, London.

CHAPTER THREE

Claude Bernard

'Scientific experiment, tracing facts to their origins, must in turn enlighten our minds, refine our sensitivity and strengthen our intellect'

CLAUDE BERNARD, INAUGURAL ADDRESS TO THE ACADÉMIE FRANÇAISE IN 1869

The experiments of Claude Bernard (Fig 9) are central to the story of curare. They provided an essential link between the observation made by Benjamin Brodie and Charles Waterton – that curare killed its victims by paralysing the muscles involved in breathing – and the analysis of the way it acts on the muscle itself. In his experiments, Claude Bernard demonstrated the way in which curare caused the breathing to stop and the limbs to become paralysed, and how it caused the death of the animal. It is possible that, had Bernard not decided to experiment with the gift of a sample of the poison, casually given to him by a friend, it would have remained a pharmacological curiosity.

Claude Bernard was a remarkable scientist. He predicted many of the findings about the way the body works long before the tools were available to allow his theories to be tested and verified. As a result, the credit for many of his ideas and innovations ended up being attributed to others. He is central to this story because he demonstrated where curare acted when it paralysed its victim. He is also important for his advocacy of scientific methodology. In his 1865 book on the scientific method, *An Introduction to the Study of Scientific Medicine*, he wrote, 'We formulate a hypothesis elucidating the relationship between cause and effect for a particular phenomena. We test the hypothesis. When the hypothesis is proved it becomes a scientific theory.'

He was ahead of his times in his insistence on testing and validating new ideas by experimentation. He was prepared to challenge preconceived beliefs and the rigid hand of the orthodoxy of scientific authority; he was an iconoclast. He wrote, 'One must not uncritically accept the authority of academic or scholastic sources.' Just because a scientific celebrity or a particular group of researchers held a certain view, that did not make it a fact. To separate fact from belief, one needed experimental evidence.

This led him to assert that 'when we meet a fact which contradicts a prevailing theory we must accept the fact and abandon the theory'. An attitude that today we associate with the twentieth-century philosopher, Karl Popper.

Possibly Bernard's most famous hypothesis, which he proposed long before it was rediscovered by the American scientist Professor Walter Bradford Cannon, who called it *homeostasis*, was his assertion that the purpose of all biological activity is to minimise any fluctuation in the composition and temperature of the body fluids.

'*La fixité du milieu intérieur est la condition d'une vie libre et indépendente*' ('The constancy of the internal environment is the condition for a free and independent life') – meaning this process was essential for what he called 'a free and independent life', life that could be sustained in the face of changes in the environment.

The intellectual whirlwind in Europe, created by the Age of Enlightenment, blew away many of the old beliefs, paving the way for new ideas. Nowhere was the effect more keenly felt than in France. In the nineteenth century the unquestioning attitude to religion, the nature of society and the place of commerce was, for the first time, openly discussed and debated. The rash of pamphlets and publications that appeared on the streets of the great cities produced a climate of change that culminated in the visions of a new society, as espoused by Victor Hugo, Émil Zola and Karl Marx.

It also produced an explosion of interest in science and nature. In French science, it was the time of Antoine Lavoisier, Joseph Gay-Lussac of the Marquis de Laplace, and of Buffon and Lamark. It was marked by a burgeoning of laboratories, museums and scientific publishing. The previous ready acceptance of the established dogma on which much of science had rested was challenged by new information and original ideas. Thanks to the efforts of scientists such as Bernard, the teaching of science became subject to the rigors of objectivity and experimentation.

All over Europe, inductive reasoning – the testing of ideas by experiment and arguing from specific evidence to general hypotheses – replaced teaching by deduction and decree. In England the establishment of the Royal Institution in London at the end of the eighteenth century had provided laboratories for Humphrey Davy to work on oxygen and nitrous oxide, work he had started at the Pneumatic Institute in Bristol; and for Michael Faraday to perform his experiments on electromagnetism. The institute provided a focus for the advancement and dissemination of knowledge in the physical sciences.

The nearby Royal Society flourished as a centre of debate on the mysteries of natural science with Thomas Huxley, Joseph Banks and Charles Darwin often contributing to their proceedings. In medicine, for the first time, a licensing examination was introduced for those who wished to practise as doctors, and the great medical schools of London and Edinburgh were established. In Britain, John Hunter, Joseph Lister and John Snow introduced scientific method, based on experimentation and observation, to the teaching of medicine. Museums were built, new scientific journals were published, continents were explored and universities, free of religious control and political restrictions, were established.

These developments were echoed on the continent of Europe. In Leiden in Holland, Hermann Boerhaave's teachings had laid the foundations of a new scientific tradition based on measurement and observation. It was in Leiden that the naturalist Carolus Linnaeus proposed his classification of plants, and Franciscus Sylvius described the anatomy of the brain.

In Germany, Austria and Hungary important centres of learning were established. These produced original thinkers such as Robert Koch, Rudolph Virchow and Ignaz Semmelweiss, whose work on infection and disease changed the face of medicine and opened the door to the development of surgery. In France, the Curies, Marie and Pierre, demonstrated the powers of radon, and Louis Pasteur showed the part played by microorganisms in fermentation and disease. The nineteenth century was a golden age for science.

The Collège de France was one of the main teaching institutions in Paris. It was an unusual university with a long history of challenging convention dating back to its foundation by Cardinal Richelieu. The staff, in the form of the Assembly of Professors, elected new professors not according to the subject they taught, but as a result of the current *importance* of what they taught. As a result, a professor of Greek would be likely to be succeeded by one who taught philosophy. There were no compulsory studies, tests or examinations. It was here that Frédéric

Joliot-Curie lectured on physics while Michelet espoused the virtues of the new liberalism.

In the early part of the nineteenth century Dr Françoise Magendie, a physician and surgeon to the Hôpital Dieu in Paris and one of the pioneers of experimental physiology, established a school teaching physiology and anatomy at the college. It was based on what he termed 'repeated and reproducible experimental proof'. He would tell his pupils, 'I only have eyes, not ears,' as he instructed them to believe only what could be tested and observed.

By all accounts Magendie was an eccentric and difficult colleague. He was sarcastic, autocratic and slow to give praise. He is reported as declaring, in one of his lectures, that, if the doctors were driven from the hospital, the death rate might drop. He seldom acknowledged the contributions of his co-workers. This led to his prolonged enmity and rivalry with the distinguished experimental Parisian physiologist Marie François Bichat. Magendie is remembered today as the person who demonstrated the way in which the nerve roots carrying sensation entered the spinal cord.

In 1839, a tall, good-looking, shy twenty-six-year old student applied to become his unpaid assistant so that he could work for his doctorate in medicine. It was the start of the extraordinary career of Claude Bernard that was to bring him great honour and fame.

Claude Bernard was born in 1813 in the small country town of Saint-Julien, near Villefranche-sur-Saône, in the Rhône Valley, just north of Lyon. His house, to which he would frequently return in his later life, looked out onto the hills of Beaujolais and the valley of the Saône. His father, who owned a small but successful vineyard died when he was young and Claude was brought up by his mother and educated by the local Jesuit priests at the Collège de Thoissey, where his lessons comprised the teaching of Latin, Greek, arithmetic and geometry. He is said to have been an average scholar, finding Latin particularly tedious.

Although it was his mother's hope that he would take over the family business when he left school, Claude was reluctant to commit himself to a rural life. After a year or two in the vineyard, he became restless. Even in the remoteness of the country he must have been aware of the changes that were occurring in France. This was a time when literature, science and new art forms were flourishing in Paris. There was a restlessness within society, where the poor lived in wretched conditions while all around were the trappings of great wealth. It was fanned by the liberal writings of the great French authors, such as Victor Hugo, who risked banishment for promulgating their revolutionary views. All of this must have been evident to the young Claude Bernard, even in the tranquil hills of the Rhône Valley.

At the age of eighteen, Claude left the vineyard and travelled to the nearby town of Lyon, where he found employment as an apprentice to a pharmacist, Monsieur Miellet. He lived over Miellet's shop doing such menial tasks as sweeping the shop and washing the bottles used for medicines. He progressed to delivering the various potions to patients and to making up medicines to simple formulae. He does not appear to have found the employment particularly absorbing, and, from remarks he made in later life, he seems to have had a low regard for the efficacy of the remedies that were being dispensed.

He was more interested in the theatre than the pharmacy, and during his spare time he wrote a light-hearted comedy, *La Rose du Lyon*, which was performed at the Lyon Theatre. It ran for one season to less than enthusiastic acclaim. Undeterred, Claude wrote a second play, a prose drama in five acts, *Arthur of Brittany*. Unfortunately, no copy of the play has survived, and we do not know the nature of this composition.

However, it is recorded that, following up a casual introduction, he took the play to the distinguished literary critic Saint-Marc Girardin, in Paris, for his opinion. To someone usually described as shy by nature, this must have taken considerable courage. Girardin's criticism was short and to the point, 'Forgo your literary ambitions,' he replied. 'For the benefit of science become a doctor.' To the twenty-one-year-old Claude, this scathing criticism from one of the most distinguished men of letters of his time was sufficient to end his ambitions as an author. Whether or not it was as a result of Girardin's criticism or from his observations working in the pharmacy in Lyon we do not know, but Claude decided to go to Paris to study medicine.

Claude Bernard and medicine

Claude enrolled at the Paris School of Medicine in December 1834. Although his mother sent him money for his lodgings in Paris, he was desperately hard up. In order to help pay his expenses, he took to teaching at a local girls' school and giving private coaching in the evenings. This was a difficult time for Claude. He does not appear to have been a particularly good student, although he regularly attended the anatomy classes (it is known that at one of these classes he met the composer Hector Berlioz, who at that time was also studying medicine).

At the end of his studies he took the Interne des Hôpitaux de Paris examination and just scraped through, passing twenty-sixth out of twenty-nine pupils. It was after this experience that he approached Professor Magendie at the Hôpital Dieu. Magendie had established a reputation as a physician and a talented research worker. Bernard asked permission to become his pupil and research assistant at the Collège de France so that he could study for his doctorate.

Magendie's principal interest, at that time, lay in the way that food was digested in the gut. There were two different widely held beliefs on the subject. Some thought that the glands of the digestive system secreted only one digestive juice, whose activity changed according to whether it was in the acid of the stomach or the alkaline media of the intestine; others maintained that different juices were secreted in separate parts of the digestive system in order to achieve different effects.

It was natural, therefore, that Claude Bernard should start his pupillage by investigating the actions of various enzymes on the digestion of foods in the gut. He was thirty-one when he published his thesis on the role of pancreatic secretions in the digestion of fats. He not only described the role of pancreatic secretions but it is evident from his writings that he studied the passage of fat into the bloodstream.

It was at this time that he formulated the concept of internal and external glandular secretions, which he presented in his second paper on the workings of the pancreas. He suggested that there are two different types of gland in the body. One set of glands, such as those of the pancreas, stomach and salivary glands, produce secretions such as saliva and gastric juice that are excreted outside the body or into the digestive system. He called their juices 'external secretions'. The secretions of the other glands are invisible as they pass directly into the bloodstream. The products of these he termed 'internal secretions'. He believe the pancreas not only produced the external secretion – the pancreatic juice – but also secreted another substance, which went directly into the bloodstream, which influenced the uptake of glucose by the liver.

Bernard was appointed *préparateur* (assistant) to Magendie at the Collège de France in 1844. As a *préparateur*, he was able to perform his own experiments, assist the great man in his demonstrations and to stand in for him when he was unable to give his lectures. In his reference for the post, Magendie grudgingly admitted

Bernard's skills as an experimenter and dissectionist. He taught Bernard the discipline of research and the necessity of verifying his results by experiments in more than one animal species. He impressed on Bernard the need to test any hypothesis by experiment before accepting its veracity. It is clear, from his writings on scientific methods, that Bernard took his tutor's words to heart.

In 1845 Claude Bernard married Marie (Fanny) Martin, the wealthy daughter of a doyen of the Paris establishment. The marriage was arranged by one of his colleagues. It is clear that, in both their backgrounds and their aspirations, they were an ill-matched couple. Although his wife's dowry brought some financial relief to the hard-pressed scientist, it does not appear to have been more than a marriage of convenience.

When not required for his duties at the college, he started working at a laboratory that had been set up by his friend, the revolutionary chemist Jules Pelouze. The accommodation at the laboratory was primitive; it was little more than a shed that had once functioned as a stable in the run down area of Paris known as the Coeur de Commerce Saint-André. Here, in a smelly semi-basement with a leaky roof, Bernard set about performing his famous experiments on the role of the liver and pancreas in maintaining the level of sugar in the blood.

It was these experiments that led him to conclude that, following a meal, especially one containing a lot of sugar, the liver stored excess sugar in the form of a substance he termed *glycogen*. He demonstrated that the liver released the sugar from glycogen when the blood reaching it was starved of carbohydrate.

In his second paper on the functions of the pancreas, he postulated that this process was controlled by an 'internal secretion'. In these experiments he predicted the role of insulin over sixty years before it was finally demonstrated by the physiologist Edward Sharpey-Shafer in England, and sixty-four years before it was isolated by the medical scientist Frederick Banting and his assistant Charles Best in Canada. It was for this work that Bernard earned the sobriquet of 'father of endocrinology'.

Among the other experiments that he conducted at this time was the demonstration that the concept of body heat proposed by Lavoisier was wrong. It was a brave thing for a budding scientist, such as Bernard, to suggest that Antoine Lavoisier, who was an icon of French science, was wrong. It was Lavoisier, together with Joseph Priestly in England, who had demonstrated the presence of the new gas, oxygen, in air and his opinion was regarded reverentially by the scientific establishment.

Lavoisier had taught that the body produced its heat in the lungs by burning the sugar contained in blood in the presence of oxygen taken in during respiration. In order to test this hypothesis, Bernard, no doubt encouraged by his various collaborators, including Becquerel and Magendie, inserted heat-sensitive probes into the main blood vessels in the necks of animals and manipulated them along these vessels into the heart. He placed one into the right side of the heart, where the blood that had been returned from the body was pumped into the lungs. The other probe was passed into the left side of heart, where the temperature of the blood coming from the lungs back into the heart was recorded. This allowed him to determine whether any heat had been generated during the passage of blood through the lungs.

In his notes he gives detailed instruction as to how the catheters were to be advanced so as not to snag the valves as they entered the heart. The technique he used is very similar to that used about a hundred years later by Andre Cournand, in America, to study cardiac physiology. Today, the same technique, reintroduced some hundred and forty years ago, is used to measure the activity of the heart in cardiac surgery and in intensive-care units. By comparing the temperature in the blood passing his two heat-sensitive probes, Bernard was able to show that the lungs were not a source of heat production, as there was little difference between the temperature of the blood entering the lungs and that leaving them.

It was during this period that Bernard formulated one of the most important principles in physiology. He proposed that the purpose of all the various complex biological processes that occur in the body are aimed at one end, that of keeping the composition of the body fluids within the very narrow parameters essential for sustaining life. He formulated the concept of a 'constant internal environment'. He reasoned that any significant change in this 'internal environment was incompatible with normal existence'; 'all the vital mechanisms, varied as they are, have only one object, that of maintaining constant the conditions for life in the internal environment.'

He reasoned that the brain acted as a control centre for all these systems of the body and by its nervous activity it adjusted the internal and external secretions in order to maintain the optimum body temperature, the concentration of glucose in the blood, the concentration of salt and water in the body fluids, and the acidity of the tissue fluids, at the level necessary for life. This truism was rediscovered in the twentieth century by the Harvard physician Cannon, who coined the term *homeostasis* for the phenomenon.

From this time on, the concept of the constant internal environment was the central idea that became the driving force behind Bernard's investigations. He wanted to unravel the way in which the body maintained the equilibrium in the blood and body fluids, what he termed the internal environment. He hit on the idea of using poisons as a forensic tool in the expectation that, by observing the effect of poisoning one particular system at a time, he could understand the part it played in maintaining the status of the internal environment. He wrote, 'Toxins are physiological instruments, more sensitive than mechanical means in dissecting the anatomical elements of living organisms.'

Claude Bernard and curare

It is evident that Bernard's experiments were influenced by the availability of experimental animals, analytical techniques and the very latest gadgets and measuring devices required for his experiments. For example, he acknowledges his indebtedness to Ms. Pulvermacher of Paris for making the galvanic (electric) stimulators, in the form of tweezers that he used in many of his experiments. (Fig 10) He was fortunate to have as colleagues Dr Jules Pelouze and Charles Barreswil, both distinguished chemists who performed many of the chemical analyses that were necessary for his work. It was Pelouze who brought the specimen of curare to the laboratory in the Coeur de Commerce in 1844. He had been given the poison, together with several poisoned arrows, by a friend, Dr Goudot, who had recently returned from a trip to Brazil. It was the chance meeting between Goudot and Pelouze that provided the specimen of curare that was to play such an important role in our understanding of the workings of the body.

Bernard was not a pharmacologist and it is evident that his initial experiments with the curare he had been given were of a casual nature. It was some months before he returned to them, and it was two years before he thought the findings worthy of reporting to his peers.

Nevertheless, it was as a result of the experiments that Claude Bernard carried out with this specimen of curare, between 1844 and 1856, that the way in which it paralyses and kills its victims was unravelled.

One can envisage this tall, good-looking man with his receding hairline, dressed in a frock coat with a scarf around his neck, often wearing a top hat to protect him from the rain seeping through the leaking roof, totally absorbed in his experiments and oblivious of all that was going on around him. Outside the dimly lit, smelly, damp laboratory the world was in chaos. This was the time of revolution and the Paris Commune, and of Karl Marx and his Communist Manifesto. He could not have been completely isolated from these events as his companion, Dr Pelouze, was an active revolutionary.

It was during this time that his marriage started to fail. His wife, who had helped him during his early days when he was dogged by financial problems, was strongly opposed to vivisection. It would have been surprising if any marriage could have survived Bernard's single-minded determination 'to unravel the laws that govern physiological phenomena', and it is little wonder that little is heard of his wife and daughters after this time, although it is reported that they became active anti-vivisectionists.

Bernard's first experiment with curare was carried out in 1844 on a rabbit, which he impaled with an arrow poisoned with the curare. He records in his notebooks that in five minutes the rabbit was quiet and fell over. It lay still while its heart continued beating. He noted that, although the rabbit was clearly still alive, there was no response to pinching its skin. An autopsy revealed no apparent cause of death. In the next experiment, carried out some months later, he fed a rabbit with 10 mg of the crude curare and showed that it had no effect; it was not poisonous if taken by mouth. What is surprising is that, by performing these experiments, Bernard demonstrated that he was either unaware of, or did not believe, the results of previous experiments carried out in Leiden by the Abbé Fontana and by Dr Brockelsby, who had earlier demonstrated the effect that he found. Perhaps he was following his own dictum of never believing anything unless he had been able to verify it by experimentation.

Bernard's experiments are most cogently presented in his reasoned accounts presented in his *Leçons sur les effets des substances toxiques et médicamenteuses*. ('Lessons on the effects of toxic substances and of medicaments.') These are the reports of the series of demonstrations he gave before groups of twelve to twenty students at the Collège de France after he was established as a professor in 1854. (Fig 11) However, the recent publication of translations of his laboratory notebooks makes it evident that these events were largely stage-managed and were invariably based on studies he had conducted years earlier in his laboratory in the Coeur de Commerce.

In this manner, the observation made in an early laboratory experiment – that curare did not stop the heart beating – became one of his set-piece '*leçons*'. In the public demonstration, he inserted a manometer into the carotid artery so that all could see that the heart went on beating and there was little change in the blood pressure for up to twenty minutes, in spite of the unresponsive state of the animal. His notebooks reveal a less organised progression in the experiments carried out in his laboratory than in those presented to his distinguished audience during the formal lessons.

There is a gap of some years between some experiments, presumably because it was difficult, in the turbulent times of the 1850s, to obtain sufficient curare and experimental animals to complete his investigations. However, his notebooks illustrate his uncompromising attention to detail in all the studies. Following the teachings of his mentor, he invariably repeated all the important experiments on three or four occasions, usually using different animals for each investigation. Many of the studies contain sketches and diagrams to illustrate important details.

Working with the original Pelouze specimen of curare in 1844, Bernard records that he introduced a fragment under the skin of a frog, which caused it to become paralysed. Using his new stimulating galvanic tweezers, he observed that, if he cut through the skin and stimulated the muscle under it directly, it caused the muscle to contract. However, it failed to respond when the nerve, rather than the muscle, was stimulated.

In unpoisoned animals the muscles can be made to contract, either by stimulating them on their surface – so-called 'direct stimulation' – or 'indirectly', by stimulating its nerve some distance from the muscle. By demonstrating that the muscle was capable of reacting to direct electrical stimulation but not to stimulation via its nerve, Bernard demonstrated that the poison acted on the nerve, not the muscles.

In the next experiment, carried out some six months later, he introduced curare under the skin of the upper part of the body of a frog and recorded that, when the animal became paralysed, stimulation of the skin over the forelimbs caused the hindlimbs to move. Bernard interpreted this as evidence that the nerves carrying sensation to the brain and spinal cord are unaffected by curare, while the motor ones, those causing muscle contraction, are poisoned.

His attempt to demonstrate the exact site of action of the poison was less than convincing. He dissected out the muscles of the two legs of a frog together with their nerve supply and removed both specimens from the animal. He immersed the nerve to one muscle in a watch glass full of saline containing curare, taking care to leave the muscle outside. He placed the other muscle in separate bath of curare while the nerve was left out of the solution. He noted that the specimen with the immersed nerve still responded to stimulation but that the one whose muscle belly, together with the very end of its supplying nerve, was curarised did not respond. He interpreted this as evidence that it was the nerve that was poisoned by curare. 'Poisoning by curare', he asserted, 'kills the motor nerve even though the poisoning starts at the peripheral end.'

Although the experiment showed that curare acts on the very ends of the nerve, where it actually enters the muscle, the experiment failed to show an effect on the nerve trunk itself.

In 1855 Bernard was appointed to the Chair of Physiology at the Collège de France in succession to Magendie. In the following year he started his famous course of lectures at the *collège* on poisons and toxic substances These were eventually collected together and published. The logic presented by the progression of ideas in these simple, clear-cut experiments gives the reader enormous confidence in their conclusions. Their simplicity belies the painstaking preparation and research that made them possible.

His lectures were so popular that people came from far and wide to see the great man in action. Émil Zola, Paul Bert, Charles-Édouard Brown-Séquard, the neurologist who succeeded Bernard at the Collège de France, and Louis Pasteur are believed to have visited the college to hear Bernard. Indeed, it is probable that the character Dr Pascal in Zola's saga *Les Rougon-Macquart* is based on Bernard.

Bernard's *leçons* started in May 1856 and continued over the following ten years until his illness in 1866, but it is Lessons 22 to 25 that best illustrate the convincing nature of his work with curare.

On 28 May 1856 he opened the first lesson by reminding his pupils that, when studying the effect of a toxin, they must consider all the possible systems it might affect. The first task confronting the experimenter is to narrow the field down by excluding those that are not affected. In this way the particular site at which the poison works might be identified for study. He remarked, 'Gentlemen, in order to study the action of curare, like that of any active agent, it is necessary to examine the changes that occur under its influence in various systems: circulatory, nervous, glandular, et cetera. We will begin with the influence of curare on the blood and circulation.'

The initial demonstrations were designed to exclude those biological systems that were unaffected by the poison. He showed that curare did not kill by causing heart failure, and it did not poison the blood. On 28 May, he demonstrated that curare added to blood in a test tube did not interfere with its 'essential property', the uptake of oxygen or the 'throwing off of carbonic acid'. In the same lesson he demonstrated that curare did cause a small fall in blood pressure in a dog but that the fall was 'far from being in proportion to the animal's state of apparent death'.

Bernard contrasted the effect of curare with that of other poisons such as strychnine in frogs. He observed that, unlike most poisons that cause convulsions and nervous overexcitement, curare 'abolishes the properties of the nervous system associated with hyperactivity' and that this results in the paralysis that is ultimately the cause of death.

Bernard's declared aim in studying the effects of poisons such as curare was to understand how the body worked. In the public experiments described in Lessons 22 to 24 he showed that muscles continued to be able to contract normally in response to an electric shock, even when the nerve supplying them was poisoned. The demonstration involved inserting a small dose of curare under the skin of the frog's back. After about three minutes, he records, the frog lay inert and apparently lifeless. Nevertheless when the muscle of the leg was exposed and 'galvanised' (electrically stimulated), it still contracted but, if the nerve supplying the muscle was stimulated in the same way, nothing happened. It is clear that curare did not poison the muscles, but caused paralysis by preventing the nerve from working properly. This confirmed Bernard's belief that curare acted by poisoning the nerves that supply muscles.

In a subsequent demonstration he compared the effect of curare on a frog with that of decapitation. He showed that there was a difference in the way the two frogs responded to stimulation of their nerves. In the animal that had received curare, electrical stimulation of the nerves produced no effect. In the decapitated one, stimulation of the nerve caused contraction of the muscle it supplied. These experiments showed that curare did not poison the brain – had it done so, the effect would have been similar in both the decapitated and curarised animals – but its action was restricted to the nerves supplying muscles. Bernard summed this up as demonstrating that curare poisoned nervous irritability but not muscle contractility.

As a result of these experiments, Bernard was convinced that curare prevented the messages carried in nerves from causing a contraction in the muscles they supplied. It was now necessary to find out whether curare affected all nerves or just those supplying muscle. To separate out the motor nerves, the ones supplying the muscles, from those carrying information from the skin and other sensory organs presented a problem. The only way he could be sure that the sensory nerves were functioning normally was to demonstrate that a stimulus, such as pinching the skin, produced some form of mechanical response. The problem was that, in order to see any movement in response to stimulation, the animal had to be able to contract its muscles. If he poisoned the animal with curare it would be unable to respond.

Bernard met this difficulty in the ingenious experiments carried out on 4 and 6 December 1854. In the first of these he tied the blood vessels supplying the left hindlimb of a frog, isolating it from the general circulation. He then proceeded to poison the frog with a small dose of curare. In fifteen minutes the frog lay still and paralysed. After a few minutes he pinched the skin and demonstrated that only the left limb, which had been isolated from the circulating curare by the ligature around its blood supply, responded; the other curarised limbs lay paralysed.

This effect was positively confirmed in the next demonstration, which was carried out in December. In this he totally isolated the frog's legs from the rest of the body leaving only its nerve supply intact. Twenty minutes after giving the frog a dose of curare, he showed that pinching the skin of the animal caused movement in the limb in which the circulation had been interrupted and whose nerve had been isolated and therefore protected from the effect of the curare in the blood.

These experiments showed that nerves carrying sensory messages to the frog's brain are unaffected by curare in a dose that paralyses the animal. It also demonstrated that the spinal cord was unaffected by curare, as it was able to relay the sensation produced by pinching the skin to the spinal cord and from there to the muscles, even though only those muscles unaffected by curare could respond.

These two experiments led Bernard to conclude that curare was carried by the bloodstream and that it poisoned only the nerves supplying muscles. It did not poison the nerves that carried sensation from the skin to the spinal cord and brain.

His conclusions are stated in Lesson 22 given on 30 May 1856: 'Curare acts on the nervous system but it must not be thought that it acts, simultaneously and in the same way, on the sensory and motor properties. Its transient effect excludes the idea of an anatomical lesion…'

By means of this series of elegant experiments, Claude Bernard demonstrated that curare prevents the messages carried in the motor nerves from reaching the muscles. He concluded that it selectively and reversibly poisoned the nerves. He was almost right.

After Bernard

At the time of Claude Bernard's experiments it was uncertain whether there was anatomical continuity between the nerves and the muscles, or whether there was a gap at the nerve ending. At about this time the German optical company of Carl Zeiss used the new optical grinding techniques to produce superb magnifying lenses. When these were used in the construction of microscopes they greatly enhanced the magnification that was possible.

It was with the help of these microscopes that Virchow, at the Charitie Hospital in Berlin, demonstrated that the organs of the body are made up of discrete cells, a discovery that allowed him to establish the discipline of histology.

In 1862, Professor Wilhelm Kühne, with the help of one of these new microscopes, was able to demonstrate a slight swelling at the end of the motor nerve, which he termed the *nerve bulb*, and to show that there was a gap that separated it from the muscle. He also recognised that the muscle surface opposite the swollen end of the nerve differed from that of the rest of the muscle. He termed this region between the nerve and the muscle, the *end plate* (Fig 12). It was Bernard's pupil Alfred Vulpian who concluded, in his thesis in 1866, that this was the likely site of action of curare. He reported that, when curare was applied close to the end plate, it caused rapid profound paralysis, whereas its application directly to the motor nerve near its end produced a slower and lesser effect. He concluded that 'curare interrupts the communication between nerve fibres and muscle fibres'. We now know that Vulpian was right.

On 27 May 1868, at the Palais Mazarin in Paris, Claude Bernard was awarded France's greatest honour: he was elected to be one of the selected *quarantes*, the forty most distinguished men of science and letters who constituted the Académie Française. As if by fate, one of his fellow academicians on that day was Saint-Marc Girardin, who all those years ago had advised Bernard to give up trying to write plays and instead to become a doctor.

In his latter years Bernard became increasingly obdurate. He was unconvinced by Pasteur's theory of micro-organisms as the cause of fermentation and he failed to give credence to Vulpian's evidence that curare acted at the end-plate junction between the nerve and muscle. He spent much of his time in the family house where he was born, in Saint-Julien in the Rhône Valley. He died peacefully in his sleep in 1878. In spite of his fame, Bernard died a lonely man; before his death he tried to become reconciled to his wife and his two children, but his efforts were in vain.

Claude Bernard was one of the great men of science but his genius is seldom recognised outside France. Perhaps this is due to the successes of other eminent scientists at this time, such as Pasteur, Koch, Lister and the Curies, whose research bore fruit immediately. All the ideas promulgated by Bernard had to wait many years before their importance was recognised. The importance of his deduction about the role of the brain in maintaining the stability of the internal environment of the body waited almost a century and a half before it was rediscovered. His predictions about the internal secretion of the pancreas and the part it played in controlling the level of glucose in the blood did not become recognised as true until some sixty years after his paper on the subject.

Bernard was a man before his time. He lived before techniques were invented that would have allowed him to prove his various hypotheses. Even though he unravelled the way curare paralysed its victim, the microscope that might have allowed him to identify the actual site of action of the poison did not come into use until after he had retired.

Although today Claude Bernard's contribution to medicine is scarcely remembered outside France, such was his fame when he died that he was given a state funeral. He was buried in the Lachaise Cemetery in Paris.

CHAPTER FOUR

Dangerous Remedies

News of Claude Bernard's experiments soon spread across the Channel to England. His observations on curare were published in his famous book, *Leçons sur les effets des substances toxiques et médicamenteuses* in 1856. It was translated into English within two years.

The minutes of the Royal Society in London show that the results of his experiments, which were initially regarded with considerable scepticism, were nevertheless frequently discussed in the meetings of that august body. Nevertheless, many English savants were reluctant to accept the findings of foreign, especially French, scientists. The disagreements between Priestly and Lavoisier as to the nature of the recently discovered gas, oxygen, had resulted in an atmosphere of mutual suspicion between the scientists in the two countries.

The surgeon Thomas Spencer Wells was particularly outspoken. He belittled the work of Claude Bernard in public and refused to believe that a foreign doctor, such as the Italian, Dr Vella, had used curare successfully to treat convulsions in patients in Paris. In this he was probably right! However, it was possible that it was Dr Vella's report that put the idea of using curare to treat the convulsions of tetanus into his mind.

Rabies and tetanus

Victorian England saw an Industrial Revolution and with it a major shift of the population from the countryside into the towns. London in particular had become a huge metropolis. These demographic changes brought with them a change in the pattern of disease. The demands of a densely populated city necessitated the introduction of many major public-health measures. The completion of the major sewer system taking London's waste far down the Thames, where it would be washed out to sea and no longer contaminate the drinking water, finally put an end to the frequent outbreaks of the dreaded cholera; and the introduction of measures to reduce the rat population had made the outbreaks of bubonic plague rare.

One aspect of the rapid increase in the population was an increase in overcrowding, alcoholism and poverty. As a result, there was a surge in illnesses such as tuberculosis, venereal diseases, and contagious diseases such as smallpox, diphtheria, measles and scarlet fever. An alarming development was the increasing prevalence of two diseases that had hitherto been largely associated with the countryside: hydrophobia (rabies) and tetanus (lockjaw). Both of these diseases caused convulsions and severe, life-threatening muscle seizures.

Hydrophobia resulted from a bite by a rabid dog or fox. Although it was known that rabies was spread in the saliva of infected animals, it was not until Pasteur demonstrated the infective bacillus that the true, infective nature of the disease was appreciated.

Tetanus was caused when a wound, often quite minor in nature, became infected with the spores of the tetanus bacilli. In the first half of the nineteenth century tetanus was not recognised as an infective disease. It is clear from the discussions, in the various medical bodies at the time, that its cause was unknown. It had been suggested that there were two different types of tetanus: one occurred in the countryside, in which the convulsions were more or less continuous; the other, with intermittent seizures, was seen after surgery.

It was believed that the form of the disease found in patients recovering from surgery was due to 'chills'. In one patient who suffered the disease after surgery it is reported as being the result of the 'draught from a ventilation shaft near the patient's bed in which the boards had become dislodged'. It was only recognised that spores were the true cause of the disease when Pasteur showed that, after lying dormant for months or years, they could germinate when the conditions were favourable and cause infections.

The spores of tetanus are very resistant to attempts to destroy them. They often lie, undamaged and potentially infective, in the soil for years. They can withstand freezing conditions and heat waves without any effect on their capacity to cause disease. Spores, taken in with the grass eaten by ruminant animals such as sheep, cows and horses, are excreted in their faeces. It was the spores excreted in the faeces of animals that caused tetanus in humans when they contaminated a wound.

The increase in the incidence of cases of tetanus in Victorian London was particularly associated with an increase in the number of horses in the capital. This was the result of the rapid expansion in horse-drawn transportation. Numerous horse-drawn wagons and hansom cabs were required to service industry and the burgeoning urban population. Behind the rows of elegant new Victorian houses were the 'mews', where the horses and carriages of the wealthy were kept while the hostler and his family lived in the loft space above. It was common for the poor to collect horse manure from the mews and from the streets for use as fertiliser or, when dried out, as a fuel.

With the passing of the New Poor Law Act in 1834, many indigent Londoners were confined to the new poor law buildings where they would be given a plot of land, an allotment, to cultivate in order to provide themselves with staple foods. This was the start of the allotment system that survives to present times. If one of these unfortunate individuals injured himself while attending an allotment that had been fertilised with horse manure, it was likely the wound would become infected with dormant spores. Within one to six days he would develop a fever, followed by convulsions, which were likely to increase in severity until the whole body became rigid and breathing stopped.

The spasms usually started in the muscles of the face and neck, causing the characteristic 'lockjaw' and the '*risus sardonicus*' (sardonic grin) so typical of the disease. Spasm of the neck muscles would cause them to become stiff, lifting the patient's shoulders from the bed as if he were lying on a curved board. The more rapid the onset of the symptoms the more severe their disease was likely to be. Death was usually due to a combination of toxaemia from the infection and anoxia (lack of oxygen) from an inability to breathe due to spasms of the respiratory muscles. The fever and muscle activity increased the body's demand for oxygen manyfold, while the spasms of the muscles of respiration made breathing impossible. The patients invariably turned blue from anoxia before dying of heart failure.

Therapeutic use of curare

Before the advent of curare, convulsions were invariably treated with increasing doses of laudanum (opium). Although it often ameliorated the symptoms, it also depressed breathing and must have contributed to the subsequent anoxia that produced the terminal heart failure. The possibility of using the paralysing properties of curare to treat the convulsions associated with these diseases was an attractive idea that appears to have occurred to doctors in many countries at much the same time. One of those who saw the possibility of using it in this way was Squire Charles Waterton.

It is clear from the records that he played an important role in encouraging its use in England. After receiving a letter from a Dr Sibson of Nottingham, requesting a supply of curare in order to treat the spasms of tetanus, he discussed the matter with his friend Professor Sewell of the Veterinary College in London. Waterton reported that, from his discussion with Sewell, he was hopeful that curare might be useful in the treatment of tetanus and hydrophobia.

In 1838–9 he undertook a special visit to Guyana 'to obtain a store of the poison with a view to its use in tetanus and hydrophobia'. He provided the poison for use in the treatment of this 'awful disease' and was a ready source of advice on how to administer the drug. Waterton was aware of the jealousy that was developing over who would gain the credit as the first person to introduce this 'new cure' and wished to distance himself from the controversy. He always maintained that that he was not responsible for introducing its use into medical practice, although he claimed that it was he who suggested that it might be used to treat hydrophobia. He wrote:

> I wish it to be clearly understood that I do not claim for myself the merit of this discovery should it prove successful. I certainly paved the way to it… by going in quest of the poison at my own expense.

Many anecdotal, unsubstantiated reports of the use of curare to treat convulsions were published in medical papers around 1850, and for a time it seemed as though it was going to prove an effective remedy for what was a very painful disease with a high death rate. Once it promised to offer a cure for these diseases there was an unseemly outbreak of jealousy over who was going to get the credit for its introduction into clinical practice. Spencer Wells was particularly outspoken on the subject. He made his views clear in the discussion following a meeting of the Medical and Chirurgical Society (now having added 'Royal' to its name, in 1834) of London on 3 December 1859 in which he echoes the general disdain of the British establishment in Victorian times for foreign medicine. He said:

> A great stir has been made abroad as to the researches of Bernard and Vella respecting the physiological action of woorara and the treatment of tetanus yet years ago Sir B. Brodie, Mr Sewell and Mr Morgan were doing all the French physiologists and surgeons were taking the credit for.

Curare in patients

Probably the earliest reported use of curare for a medicinal purpose is found in the writings of Robert Schomburgk, Alexander von Humboldt's friend and fellow explorer. Schomburgk writes that he tried its effect while suffering from rigors caused by an attack of ague (malaria) after his supply of quinine had ran out. In 1841 he records, in the pages of the *Annual Review of Natural History*:

> I took frequently of the urari [curare] in doses of about as much as I could get on the point of a knife to cure the fever of the ague. After taking it I generally felt a slight headache, but it did not relieve the fever.

He had previously seen natives swallow the drug, as he describes in the paper: 'they took it as a treatment for stomach pains without ill effect.' It is not surprising that he found the poison had no effect on the rigor or the fever, as it is known to be inactive when taken by mouth.

It was Charles Waterton who saw the need to try out curare in several patients with hydrophobia in order to see whether if it might save their lives by preventing the convulsions. He was encouraged in this view by his friend Sewell, the professor of veterinary science in London. Sewell was a man of science and, before he decided on a trial in patients, he persuaded Waterton to let him use it on animals suffering from tetanus. Waterton readily agreed to supply him with curare.

Sewell first tried it on two horses suffering from equine tetanus. Rabies does not occur in horses but they are susceptible to tetanus in much the same way as man. He gave sufficient drug to stop all the animal's spasms but he records that as a result of the ensuing paralysis it was necessary to perform artificial respiration for four hours to prevent the animals from dying. Although he did not describe how he achieved this, it could not have been an easy task on a horse, especially considering the primitive equipment available at the time.

He records that, at the end of the experiment, when the horse started to breathe again, the muscle spasms returned and one of the animals died from 'the flooding of his lungs with the contents of his stomach'. The other animal also eventually died. Undeterred by these results, Sewell was so impressed by the way the curare had put an end to the muscle spasms that he went on to use it in 1830 to treat hydrophobia in one of his colleagues at the Veterinary College when he was bitten by a rabid dog. What happened to the patient is not recorded, although it is probable that he survived, as, according to Waterton, he told Darwin's friend, Sir Joseph Banks, that if he were bitten by a rabid dog he would not hesitate in having the 'wourali poison applied'.

After hearing Professor Sewell's accounts of his experiences with curare, a Dr Sibson, a physician in Nottingham, wrote to Waterton so that he might also obtain some poison to try out in his country practice, where cases of tetanus were relatively common. It was following this request that Waterton made a special visit to Guyana to collect a larger sample of poison so that it could be used in more patients and its usefulness assessed.

Sibson records that he also tried the effect of Waterton's curare on two horses, which he deliberately poisoned with strychnine. He reported that, although both the animals stopped having fits, they eventually died from a 'paralysis of breathing'.

It is evident that there was much discussion among members of the scientific community at this time about the new treatment. There are many references to the use of curare and numerous anecdotal case reports, such as that of a Nottinghamshire police inspector named Phelps who was bitten on the nose by a mad, rabid dog in 1838. Waterton was asked by his physician, a friend of Dr Sibson, to come to Nottingham to treat him. Unfortunately, Inspector Phelps died before Waterton reached Nottingham with his curare cure.

One of the problems confronting those who used samples of the crude curare provided by Waterton was the lack of consistency in the contents of the crude resin. Not only did they differ in their purity and potency, but it was not uncommon for the crude samples to contain contaminants. One common contaminant was strychnine, which came from the South American creeper *Strychnos nux vomica*, which grows alongside *Chondodenron*, the principal source of curare. Sibson noted this problem in his experiments on horses. He commented that both the potency of the drug and nature of its effect depended on the source of the particular specimen of woorari he used.

Although he was aware of this problem, he apparently felt sufficiently confident to use the poison to treat two patients with rabies in 1837. We have no details of how he administered the drug or of the doses used, but he claimed that, as with its use in horses, it appeared to alleviate the muscle spasms for a short time but eventually the convulsions returned as the effect of the drug wore off. He reported that, in his experience, 'no satisfactory conclusion can be drawn as to the use of woorara' and rabies.

It was not only in England that curare was used to relieve the spasms of tetanus. There are accounts of its use by Dr Vella of Turin, who, according to Spencer Wells, practised medicine in Paris. He reported using it in three patients, one of whom survived. In another report from France, the well-known French surgeon, Dr Charles Chassaignac, told of how he used it successfully to treat convulsive spasms in one of his patients.[3]

Curare was also tried as a treatment for tetanus in America. A New York surgeon, Dr A. L. Sayers, spread a resin of crude curare on a linen swab and applied it to the wound of an Irish immigrant labourer suffering from tetanus. He claimed that it provided temporary relief but that the muscle spasms returned and the patient died. According to Sayers, he went on to treat other patients suffering from tetanus but there are no records of any of his patients recovering. It would have been extremely unlikely that his ministrations would have worked, as curare is virtually ineffective used in this way. Even if the swab had been soaked in pure curare, little, if any, of it would have been absorbed into the blood.

3 He described an anatomical landmark on the first rib, a nodule that can be felt in the neck, known as Chassaignac's tubercle.

Thomas Spencer Wells and tetanus (Fig 12)

In contrast to these sketchy, anecdotal accounts of the use of curare, the detailed record provided by Dr Thomas Spencer Wells of the treatment of five patients who developed tetanus after abdominal surgery gives us a much better understanding of the problems faced by these pioneers. Spencer Wells presented his results to the Royal Medical and Chirurgical Society of London in 1859.

Spencer Wells was a respected London surgeon working at the Samaritan Hospital in London's Euston Road. He also taught surgery at the Grosvenor Place School (later to be the site of the new St George's Hospital). He became president of the Royal College of Surgeons and was knighted by Queen Victoria.

The nineteenth century was a pioneering time for surgery. Until the turn of the century surgery was largely concerned with treating the effects of trauma, setting fractures, cutting for bladder stones, dilating strictures and lancing the boils produced by infection. Previously, it had been dominated by military surgeons, some of whom were not medically trained, while others doubled as barbers. It had only recently emerged as a separate medical discipline.

Few attempts to open the body cavities, such as the abdomen or the chest, had proved successful. In 1809 in Kentucky in the USA, Ephraim McDowell, a Scottish-trained doctor, performed a pioneering operation to remove a large ovarian cyst from inside the abdomen of a Mrs Todd-Crawford. This was the first recorded case of an intra-abdominal operation being deliberately planned and successfully carried out. The surgery took forty-five minutes and was performed without the benefit of anaesthesia. This was the beginning of modern surgery.

With the introduction of Lister's antiseptic techniques in 1845 and anaesthesia in 1846, surgeons become emboldened to perform more and more complex abdominal procedures. At the time that Spencer Wells practised, most surgery inside the abdomen was still restricted to the removal of large swellings and the drainage of cysts.

In 1857, a Dr Layman of Boston collected the results of three hundred cases of ovariotomy, an operation to drain or remove an ovarian cyst. In his series, the mortality was low: he recorded only one death from post-operative tetanus. In contrast, in Spencer Wells's practice, post-operative tetanus was a much-feared complication.

Spencer Wells noticed that these cases occurred in clusters. Often there would be three or four cases in quick succession followed by several months without a single new one. He was not alone in noting the sporadic nature of the outbreaks of these cases. Mr Curling of St Mark's Hospital had reported a similar pattern in his patients following operations for haemorrhoids. Like many of his fellow surgeons, he was baffled as to why sporadic outbreaks of tetanus, of two or three cases at a time, should occur in England but not in America.

With hindsight it is possible to attribute this to the use of catgut ligatures (made from the intestine of sheep) in

British surgical practice, whereas linen thread was much used in America. It was many years later, when the infective nature of tetanus was appreciated and the practice of soaking the catgut in alcohol was introduced, to kill the spores, that this complication disappeared.

Spencer Wells's name is famous in British surgical circles. It is invoked every day in British operating theatres when surgeons ask for Spencer Wells artery forceps (known as 'clips' or 'snaps' in the USA), to staunch bleeding. It is probable that, as the inventor of the nonslip forceps (the ratchet forceps that bear his name are in fact a modification of those described by the French pioneer of stomach surgery, Jules Péan) Spencer Wells would have used them more often than most of his colleagues, resulting in the use of many more cat gut ligatures in his patients. It is possible that his invention may inadvertently have been responsible for increasing the risk of tetanus in his patients.

In his report to the Royal Medical and Chirurgical Society of London, Spencer Wells reported that two out of the four cases of ovariotomy he performed in the month of October 1859 developed tetanus during their recovery. He was very distressed by this complication. He agonised as to the cause, as he was aware that it had occurred only once in the three hundred cases described in Layman's report. It was this that determined him to try the curare treatment that was being talked about in the various medical societies.

He read of the problems caused by the impurities in the

crude samples of drug. To overcome this problem he used an extract of crude woorara that had been purified, according to the recipe of Dr Boussingault (the personal physician of Simon Bolivar), to give an active ingredient 'curarina'. Spencer Wells knew enough about the natural history of the disease to know that the patients who developed their symptoms soon after the operation, which he called acute cases, stood less chance of recovery than those in whom the onset of spasms did not occur until six to ten days later.

The treatment was first tried on a patient whose symptoms appeared on the sixth day following her operation when she was well on the way to recovering from her surgery. The symptoms started with stiffness in the muscles of the face and the appearance of the typical 'sardonic grin'. At first he used the curarina in the manner described by Sayers, dissolved on a lint swab applied to the wound. This had no effect and four hours later he records that the patient had a violent spasm of the head and neck. This encouraged him to inject '20 minims of the solution into the cellular tissue over the angle of the jaw'. He goes on to describe its terrifying effect:

> The state of the patient almost immediately afterwards was very alarming. She fell backwards as if dead and the pulse and respiration stopped for several seconds… after a deep sigh and fluttering pulse she was able to sip a little brandy and soon recovered.

Unfortunately, with her recovery from the paralysis caused by the poison, the muscle spasms returned. In spite of this the severity of her symptoms diminished over the following two weeks and she ultimately recovered.

In his report Spencer Wells goes on to discuss four further cases in which he used curarina or woorari. After his first alarming experience he was more cautious in the dose he used and he wrote that he invariably injected the drug into the forearm, rather than the tissues around the head, in case he inadvertently introduced the poison directly into a vein. He found the curare usually reduced or abolished the muscle spasms for a period, although in one patient, in whom he had used the crude 'woorari resin', it had little, if any, effect. However, the spasms invariably returned once the effect of the curare had worn off. In some of the patients he found that it seemed to have increased their general weakness and may have hastened their demise, but it generally failed to affect the outcome, especially in the acute case, where the spasms started within four days of the surgery.

In his conclusions, Spencer Wells expressed the opinion that 'there was a need for further experiments to establish the best use of woorari in convulsive states'. He went on to suggest that, in spite of the poor results, it should be used as a last resort in severe cases of tetanus, as the natural history of the disease showed it to be otherwise fatal; that the purer alkaloid, curarina, should be used rather than the crude woorari with its variable potency; that experiments should be carried out on animals to find out if convulsions induced a tolerance to the alkaloid, necessitating increasing doses; and that experiments should be performed to determine the largest dose that could be given without having need of artificial respiration.

With the development of Pasteur's vaccine for the treatment of rabies and of public-health measures to eliminate the risk from rabid animals, cases of rabies became fewer and fewer. By the twentieth century there was no longer any need for curare in the treatment of hydrophobia.

The recognition of the role played by spores in the transmission of tetanus, the sterilisation of catgut by immersion in alcohol, the development of anti-tetanus vaccine and the diminishing role of horse-drawn transportation made tetanus a clinical rarity in the developed world. When it does occur, especially in its acute, rapid-onset form, the treatment is aimed at stopping the convulsions. The principles of the treatment were laid down by Dr A. E. Bennett, a pioneer in the use of curare, in 1941. He advocated giving as much curare as necessary to control the muscle spasms and assisting the breathing or controlling ventilation, if the patient became paralysed. Today, anyone unfortunate enough to become infected with tetanus is given one of the modern analogues of curare in a dose that will produce complete paralysis and prevent any muscle spasm. The ensuing respiratory paralysis is treated by artificial ventilation, often for several days, until the effect of the toxin has worn off, while the infection is treated with antibiotics

Curare in other convulsive states

Curare was also tried as a cure for the fits associated with epilepsy. In its extreme form, grand mal, the seizures occur with increasing frequency and severity culminating in the continuous spasms of *status epilepticus*. In *status epilepticus* the spasm of the muscles of respiration may be so severe that they prevent the patient from breathing adequately. In 1860, a Dr Thiercelin treated a girl of eighteen in status epilepticus with curare and he claimed that he succeeded in reducing the frequency and severity of her attacks so that she could live a relatively normal life.

In 1878, Dr Wright used a similar treatment in a Shropshire farmer whose epileptic attacks were becoming almost continuous. He records that after the curare treatment 'he was able to read the Bible for the first time in three years'. However, there are few details of the dose used and the way he administered the drug. There are also reports, such as that of Raynard West in 1930, of curare being used to control the painful muscle spasms associated with spastic diseases, but by the beginning of the twentieth century interest in the medicinal use of curare had largely died out.

However, by a strange turn of fate, it was its potential use to control the painful muscle spasms associated with nervous diseases that was responsible for the next chapter in the story of the South American arrow poison but before this could occur more had to be learnt about how Curare produced its effect.

CHAPTER FIVE

The Controversy

In 1854, Claude Bernard demonstrated that curare worked by preventing motor nerves from causing the muscles to contract. His pupil, Vulpian, pinpointed its site of action to the minute gap between the end of the nerve and the muscle, the neuromuscular junction. The problem that faced scientists in the twentieth century was, how does the message carried in the nerve jump the gap? Is it by some form of electrical conduction or by some other means? If it was caused by electricity, just how and where was the current produced?

The work of the physiologists working with John Langley in Cambridge had provided evidence that Claude Bernard's prediction that there must be a system of 'internal secretions' was correct. These internal secretions controlled the blood pressure, the heart rate and the movements of the gut, without our being conscious of their activity. They demonstrated that the system was controlled by the brain, which sent its messages down a special set of fine nerves – which they termed the *autonomic nervous system* – and that these released internal secretions (chemicals such as adrenaline and acetylcholine) at their ends. It was these chemicals that controlled the subconscious response of the body to exercise and stress.

Others had suggested that the motor nerves of the body, which controlled the response of the muscles, might also release chemicals at their endings in order to produce their effect. The findings of Otto Loewi and Henry Dale gave substance to this concept. This set the scene for the great dispute: was it an electric current or was it a chemical that transmitted the message, carried in the nerve, to the muscles of the body?

The electrical theory

The advances in science that occurred in the early part of the twentieth century produced evidence that tiny electric currents were involved in carrying the brain's signals to the muscles of the body. New-found interest in electricity led to the development of a sensitive apparatus that allowed the measurement of very small electrical currents. These early galvanometers were a masterpiece of engineering. They relied on the movement of a piece of gold leaf, at the end of a copper stanchion, to detect electrical activity. The apparatus was so delicate that it often took hours to set it up and great skill to achieve any meaningful results.

However, with this new instrument it was possible, for the first time, to measure electrical events in the heart and in the brain. The ability to measure the small electric charge as it passed through the heart led to the development of the electrocardiogram, or ECG. With further refinements, the instrumentation was soon sufficiently sensitive to make it possible to measure the electrical discharge in the brain and in the nerves. This, after all, was the age of electricity.

The theory that the conduction of a nervous stimulus to the muscle was the result of an electric charge that jumped the minute gap between the nerve ending and the muscle was not seriously questioned until the 1920s. Those who proposed the involvement of a chemical transmitter were considered to be mavericks. As supporters of the electrical theory pointed out, it was in concert with the teachings of the French philosopher René Descartes, who believed that the muscles responded to 'spirits carried by the nerves'. The short delay, in the order of microseconds, between the arrival of the electric current at the end of the nerve and the start of muscle contraction was attributed to the impulse having to build up a sufficient charge to breach the tiny gap between the nerve and the muscle.

By using large nerves from animals, such as the giant squid, it was possible to demonstrate that a wave of minute electric current passed along the nerve when it was activated. When the current reached the end of the nerve it caused a response, such as contraction of the muscle or increase in the secretion of a gland. It was this finding that led to the theory that the brain generated the electric current and that the nerves acted like electric cables to carry its messages to the rest of the body; this was 'the electrical theory of nervous activity'. The demonstration that the electrical activity always passed in one direction down a nerve, due to the valve-like action of the nerve endings, or synapses, explained why the flow of information was always in one direction, from the brain to the organ it supplied.

Charles Scott Sherrington

The most influential protagonist of this concept was Charles Scott Sherrington, who was the Holt professor of physiology at the University of Liverpool at the beginning of the twentieth century. He later became the regis professor of physiology at Oxford. It was his brilliant Silliman lecture, given at Yale University in 1904, that kick-started the theory of the electrical control of bodily activity. Under the title 'The Integrative Action of the Nervous System', he suggested that the electric energy produced in the nerves could facilitate some body activities while blocking others, boosting some bodily actions while inhibiting others. In his later work he elaborated on this by explaining how a simple reflex response, such as withdrawing one's hand from a hot surface, required the nerves to cause some muscles to contract while at the same time making the opposing muscles relax. He explained this in terms of electrical events having a positive or negative effect on the transmission of information. He postulated that this was due to the electricity either raising or lowering the resistance of the cell to the electrical charge carried by the nerve.

At the time, Sherrington's ideas were both convincing and persuasive. He was a colossus in the world of neurophysiology. His ideas dominated the scene until the middle of the twentieth century. Few physiologists doubted his idea that the nerves acted like wires carrying information, coded in an electrical signal, from the brain to the rest of the body. It was widely held that it was in response to such an electrical signal that a muscle supplied by a nerve was caused to contract.

It is a testament to his pervasive influence that, even when evidence was produced to show that the theory was based on a misinterpretation of his results, it took a further twenty years for his acolytes to concede that he had been wrong. It was sixteen years after the publication of the experimental evidence that first seriously questioned his theory that its author, Otto Loewi, received the recognition he deserved and was awarded the Nobel Prize.

Although Loewi's experiments strongly suggested that a chemical was involved in nervous conduction, in the end it was the English pharmacologist Henry Dale's experiments, showing that curare blocked the passage of the nervous impulse to the muscle, that persuaded many doubters. The final convincing evidence was the inability of those who supported the electrical theory to explain how a chemical poison such as curare could block transmission from the nerve to the muscle if it was due to electricity.

In spite of this, the debate continued for some years. It was argued, by some, that there were two types of transmission: quick, the result of an electric current; and slow, due to chemical transmission. For others it was intuitively difficult to understand how a chemical process involving the release of molecules of acetylcholine from the nerve ending, its passage across the gap between nerve and muscle, and its reaction at the muscle receptor site could take place rapidly enough to explain the tiny millisecond response time that is observed.

Although by 1934 Henry Dale's team in London were convinced that a chemical transmitter was involved, he found it hard to obtain absolutely convincing evidence. It was by using the paralysing action of curare to prevent contraction of the muscle that they were finally able to prove that it was the chemical, found in the blood, that caused the muscle to contract. By using curare to solve this particular problem and prove their theory, they revealed, as an incidental finding, how the poison produced paralysis, preparing the way for it to be used safely in patients.

CHAPTER SIX

Otto Loewi and Henry Dale

The chemical theory

Just as the ground rumbles some time before a volcano erupts, spewing out molten lava that solidifies and changes the landscape, so scientific changes are often preceded for some years by a build-up of dissident opinions. Eventually, either because of a chance happening or because of new research, a piece of the puzzle comes to light that pulls all these disparate opinions together to confirm what had been, until that time, a tentative hypothesis. When this occurs the scientific landscape is often changed for good.

So it was in the years between 1934 and 1950. During these years the rumblings that had started from an epicentre in Cambridge, at the turn of the century, finally erupted. As a result of these seismic events the landscape of human physiology was altered for all time. It was during this period that experimental evidence was published that established conclusively that a chemical, acetylcholine, was involved in passing the brain's messages from the nerves to the various organs of the body under its control.

The Cambridge connection

At the turn of the century, the Cambridge physiologists Walter Gaskell and John Langley were studying a little-known system of nerves that appeared to be both anatomically and functionally separate from the central nervous system. They found that these nerves controlled the workings of many of the important organs of the body such as the heart, blood vessels, salivary glands, endocrine glands and the gut, without our being conscious of their activities.

In 1916, Gaskell named this system of nerves the *involuntary nervous system*, emphasising the lack of conscious control over its activity. He realised that they formed part of an essential control system, coordinated by that part of the brain that developed early in the evolutionary process. It represented a primitive, subconscious control system.

One can trace the evolutionary development of the brain by following the increasing sophistication of the responses that each part controls. The most basic functions, such as breathing, the circulation of blood and the digestion of food, are coordinated from the part of the brain that was probably the first to develop: the hindbrain. The hindbrain sits just on top of the end of the spinal cord, like a cherry on the end of a stick. It is essential for our existence, yet, under normal circumstances, we are unaware of its activities. Without it we would faint every time we stood up, be unable to run for a bus or to digest a meal.

A little further up the brainstem are those areas that are associated with control of the body temperature and maintaining its salt and water concentrations. The hindbrain, together with some parts of the midbrain, prepare the body for the exertion associated with fighting and hunting and for the response to extreme changes in the environment. They are the control centres for the involuntary nervous system that Gaskell described.

Moving higher up the brainstem, we come to the regions that control hormonal responses, sexual activity and the perception of pain. At this level we are at the threshold of consciousness. The higher centres, in the cerebral cortex, are responsible for the interpretation of incoming sensory information, memory and emotion. They control our conscious responses and our deliberate, voluntary activity.

Langley was particularly interested in the way in which the nervous system, which Gaskell had described, produced its effects. He realised that the system was actually composed of two separate but complementary parts, which he termed the *sympathetic* and the *parasympathetic* autonomic nervous systems. The sympathetic nervous system heightens those responses that prepare the body for exercise and fighting, while those of the parasympathetic system produce the opposite effect, one that was more suited to eating and resting.

If the sympathetic nerves to the heart were stimulated, the heart rate quickened, whereas stimulation of the parasympathetic nerves caused it to slow. When they acted in concert, they formed a primitive but effective feedback system. Excess activity in one triggered off a compensatory response in the other. Langley recognised that it is difficult to explain how opposite effects can be produced by two sets of apparently identical nerves with the conventional idea that the nerves act like electric wires simply as conduits of electricity. He reasoned that, if each nerve carried the same form of electrical signal, it should produce the same response at the organ it supplied. The only way to account for the different effects was to invoke some method whereby each nervous system used a different code to transmit its signal.

The problem he faced was to suggest a mechanism by which this special code could work. He came up with the idea of a chemical key that is specific for each set of nerves. One key would produce a sympathetic type of response; the other would cause a parasympathetic action in the organ supplied. He proposed that there were two different chemical substances released at the nerve endings, one by the sympathetic nerves and another by the parasympathetic nerves, each producing a different effect.

The idea of a chemical acting as a triggering mechanism for a physiological response was not entirely new.

In 1877, Dubois-Reymond had suggested that chemicals, such as ammonia and lactic acid, might be involved in the activation of muscle contraction by nerves. This was a great piece of intuitive reasoning based largely on the increase in the concentration of these compounds that he found in the blood after extreme muscular exertion. Professor Dixon of King's College, London, revived the idea in 1900. Dixon was a pharmacologist and was interested in the effects of various naturally occurring poisons.

One such poison was muscarine, which is produced by the toadstool, *Amanita muscaria*. The poisonous effects of muscarine closely resemble stimulation of the vagus nerve, the largest of the autonomic nerves. Stimulation of the vagus causes a slowing of the heart rate and increased activity of the bowel. The same effects, together with copious salivation and constriction of the pupil, are also seen in muscarine poisoning. Dixon suggested that, although muscarine was poisonous in large doses, the effects seen when the vagus is activated might be due to the release of minute doses of muscarine. Dixon's proposal was not taken seriously at the time, but the seeds of a new idea had been sown.

There is evidence that others were also thinking along similar lines. About this time, neuropsychologist J. R. Elliot found that chemicals were consistently released when certain nerves were stimulated, but he could not tell whether they were an essential part of the transmission process itself, or were produced by the contracting muscle.

There is little doubt that the breakthrough that changed opinions and opened minds to the possibility of chemical transmission, as an alternative to the electrical theory, was due to Otto Loewi and Henry Dale. Their contribution to this story was recognised in 1936, when they jointly received the Nobel Prize for their work on the chemical transmission of nervous impulses.

Otto Loewi

Otto Loewi (1873–1961) was born in Frankfurt. He was the son of a well-to-do wine merchant. At first, his father was pleased to indulge his son's interest in art and science, since he was a bright boy and an industrious student. He graduated in 1891 with a degree in classics from Strasbourg University.

When he was a student his first love was painting and music. After graduating he immersed himself totally in his special interests, visiting art galleries and museums all over Germany and Austria studying the history of art. After he had travelled widely for a year or so, his parents persuaded him settle down and study for a profession. They suggested that he become a doctor.

Otto seems to have been a reluctant medical student, but he succeeded in qualifying in medicine at Strasburg University in 1896. His dissertation was influenced by his tutor, Professor Schmiedeberg, who enjoyed the reputation of being the 'father of pharmacology'. Almost as soon as he qualified, Loewi showed that he had little intention of spending his life treating patients. His only interest lay in research.

He attended a course on chemical analytical methods in Frankfurt before embarking on a career devoted to biological research. It was not long before he performed some remarkable experiments demonstrating the way in which proteins are built up from amino-acid building blocks. He described how the body selected certain amino acids and arranged them in such a way as to produce the various types of protein it required. In 1902, he went to London to work in the laboratory of the distinguished physiologist Ernest Starling.

It was there that he met Henry Dale, with whom he was later to share the Nobel Prize. He visited Cambridge, where he fell under the influence of the professor of physiology, John Newport Langley. It was largely because of Langley that he and his fellow postgraduate students, Henry Dale and Walter Fletcher, became interested in the possibility that chemicals were involved in nervous activity.

After his time at Cambridge, Loewi went back to Austria, where, for a short period, he worked in Vienna, before being appointed to the Chair of Pharmacology in Graz in 1909. Such was the influence of Dale and Fletcher that he abandoned his studies on protein metabolism and, by the time he returned to Austria, he involved himself completely in the study of the autonomic nervous system and in the chemicals that affected its actions.

In 1921 he made the breakthrough that was to have an enormous effect on our understanding of the role played by chemical transmitters in bodily functions. He demonstrated, for the first time, that a chemical was involved in passing on the information carried in a nerve.

Years later, in 1960, when he was living in America, he published his reflections about the way this came about.

He had been carrying out experiments on the effects of nervous stimulation on the excised hearts of rabbits and frogs to see how the response was altered by the presence of various drugs, such as digitalis and cocaine. In these experiments he used an experimental preparation first described by Langendorf. This involves killing the animal and taking out the heart together with the stumps of the nerves supplying it. A tube is then inserted into the heart through which water, containing oxygen and salts in the concentrations found in blood, is passed into the heart. As the fluid starts to perfuse the heart, it starts to beat. Once the heart is beating regularly, it is possible to study the effect on it of drugs by adding them to the perfusing fluid. Loewi's studies involved finding out how various drugs altered the response of the heart when its nerve supply was stimulated.

Loewi had suspected, since meeting Langley in Cambridge, that chemicals were involved in the nervous control of the heart. His studies on the beating heart had reinforced this view, but he had not found a way to demonstrate their presence. He tells of how, on the night before Easter Sunday 1921, an idea for testing the hypothesis came to him as he lay in bed trying to get to sleep. Half asleep, he sat up and wrote his brainwave down on a scrap of paper by his bed. The following morning he had only the very faintest recollection of the experiment he had dreamed up. The brilliant idea of the night before had become a phantom of his imagination. He found the scribbles he had made and settled down to decipher them. 'I got up immediately and went to the laboratory; there I performed a simple experiment on a frog heart according to the nocturnal design.' He goes on to describe the experiment: 'The hearts of two frogs were isolated, the first with its nerve intact, the second without.'

Both hearts were prepared in the normal way and perfused with oxygenated salt solution. He then stimulated the vagus nerve to the first heart. It responded, as expected, by beating more slowly. He collected the fluid, leaving this heart in a glass vessel, and then used this fluid to perfuse the second heart. To his delight, it caused the second heart to beat more slowly, just as if its vagus nerve had been stimulated. He had shown that a chemical substance was released from the ends of the vagus nerve in the first heart when it was stimulated and that this substance caused an effect on the second heart that was the same as what would have occurred if its vagus nerve had been stimulated. He had demonstrated chemical transmission from a nerve to a muscle for the first time, although in this case it was the muscle of a frog's heart.

Loewi wrote, 'The parasympathetic (vagus) nerves do not influence the heart directly but liberate from their terminals specific substances which, in turn, causes the well-known modifications of the heart function characteristic of stimulation of its nerve.'

Since this chemical had come from the end of the stimulated vagus nerve, he called it 'vagusstuff'. He went on to repeat the experiment at the Twelfth International Physiological Congress in Stockholm in 1926. In spite of this convincing demonstration, so entrenched was the electrical theory of nervous activity that it failed to change the opinion of most of those present.

In his second series of experiments he showed that stimulating the cardiac accelerator nerve (the *sympathetic supply*) caused the release of a chemical that caused the second heart to beat faster when it was added to the perfusing fluid.

Soon after Loewi's experiments it was demonstrated that vagusstuff was in fact acetylcholine – a simple compound formed from ammonia and acetic acid or vinegar.

The problem of analysing acetylcholine was that it is rapidly destroyed in blood. In 1910, Henry Dale, working in the Burroughs Wellcome laboratory in England, had shown that the destruction of acetylcholine could be prevented by the action of an extract of the ergot mould, a common fungus that infected cereal crops in wet weather. Dale showed the active substance in the mould to be an alkaloid, eserine. He correctly predicted that eserine worked by preventing the action of an enzyme in blood that normally inactivates acetylcholine.

This enzyme was demonstrated to be acetylcholine esterase, which breaks down acetylcholine into choline and acetic acid. By using eserine to prevent the rapid destruction of acetylcholine, it was possible to collect sufficient chemical for chemical analysis. Loewi had always believed that vagusstuff was acetylcholine, but at the time of his experiment he lacked the means of proving it.

Loewi was a dedicated research worker who shunned the limelight. He continued his work on the chemicals involved in nerve activity until his arrest by the Nazis when they occupied Austria in 1938. He tells of how he was writing up the results of an experiment he had completed, demonstrating that a substance he had found, which he believed was involved in sensory nerve transmission, was different from that released during motor-nerve activity, when the Gestapo burst into his laboratory and he was marched off to prison.

As a Jew, he feared that he was to be confined to prison or executed. So anxious was he that the results of his experiments should not be lost that he bribed his laboratory technician to post them to a scientific journal on his behalf. Thanks to the efforts of the international scientific community, under pressure from Henry Dale, he was eventually released from prison in Austria after he had been forced to hand over the proceeds of his Nobel Prize to the Nazis. He was later appointed to the Chair of Pharmacology at the New York University. He died on Christmas Day 1961.

Henry Dale (Fig 14)

Henry Dale (1875–1968) studied physiology and zoology at Cambridge, where he first came into contact with John Langley, the professor of physiology, who was to be an influential figure in his subsequent career. He studied medicine at St Bartholomew's Hospital and qualified as a doctor in 1903. Once he had qualified he threw himself into a career in medical research.

Dale was a very determined and hardworking individual. Although he made friends easily he was intolerant of criticism. He initially spent some time working with Professor Ernest Starling at University College, London. Starling had made his reputation as an experimental physiologist and, together with his brother-in-law, William Baylis, made many fundamental observations about the properties of heart muscle and the effect of heart failure. His findings became enshrined in 'Starling's Laws'. These laws are still invoked today as the basis for the treatment of heart failure.

In 1903, by the time Dale joined him, Starling had started work on the various chemicals produced in the body that affected the heart and other organs. It was this work that formed the basis of his Croonian Lecture in 1905.

Dale's time in London must have given him an invaluable insight into the discipline involved in research. After spending some time with Starling in London, he visited Frankfurt to observe the work being carried out there by the distinguished pharmacologist Professor Erlich. It was Erlich's work on the drug treatment of infection that ultimately led to the introduction of the antibacterial agents, the sulphonamides.

Dale returned to London in 1904 as pharmacologist at

the Burroughs Wellcome Research Laboratory. While in this post he frequently travelled to Cambridge to visit Langley. It is probable that it was during these visits that he met and befriended a German research student, Wilhelm Feldberg, who was working with Langley and who was to later persuade Dale to turn his attention to the chemicals involved in neuromuscular transmission. Dale joined the newly formed National Institute for Medical Research (later the Medical Research Council) in Hampstead in 1914 as its head of pharmacology. He became the director of the institute in 1928.

It was during his time with Starling at University College in London that he met Otto Loewi and they became firm friends. It is probable that it was Dale who introduced Loewi to his old Cambridge tutor, John Langley. Both Dale and Loewi were greatly influenced by Langley. Dale shared Langley's conviction that it was the chemicals released by the autonomic nervous system that control the physiological responses that are essential for the viability of our body. It was this interest that led him to study the role of these chemicals in the body and ultimately to demonstrate the central role of chemical transmitters in these processes.

At about this time, research had revealed a number of chemicals that were shown to be essential for maintaining normal health. The role of various chemical substances such as vitamins, essential amino acids, hormones and adrenaline, as well as acetylcholine, were investigated and the way they affected the body was being worked out.

The problem was to see how they all fitted into the overall control process by which our bodily functions are kept on an even keel.

Dale's early work centred on the chemicals extracted from the ergot fungus. It was known that ergot poisoning resulting from eating contaminated flour caused gangrene of the fingers and a form of dementia known as midsummer madness. It was believed that the effect was due to an alkaloid, ergotamine, produced by the fungus. Dale studied the effects of the various alkaloids of ergot and established that one of them, eserine, prevented the destruction of acetylcholine in the bloodstream. This property of eserine turned out to be of particular practical importance in some of his later research.

However, it was his seminal work on histamine, a chemical produced after tissue damage, that brought him international recognition. He demonstrated the role of histamine in allergic shock. He described the process of tissue sensitisation and the catastrophic fall in blood pressure that occurs when a sensitised organ is exposed to an antigen. This became known as the Dale–Schultz reaction (Schultz, working independently, also described the same phenomenon at about the same time).

Following Loewi's discovery of the action of vagusstuff, Dale became intrigued by the role of acetylcholine in nervous transmission. His enthusiasm was enhanced by the arrival in the department of Wilhelm Feldberg, whom he had first met when he was in Cambridge. Feldberg had been expelled from his Chair in Pharmacology in Berlin by the Nazis. In 1933 Dale invited him to join his department in London.

Feldberg knew England well and had spent some time in Cambridge working with Professor Langley, but when Langley died he joined Dale at the National Institute for Medical Research in London, for a short time, before returning to Berlin to take up the Chair of Pharmacology. He remained in Berlin until he was expelled by the Nazis.

While in Berlin, Feldberg had investigated the possibility that Loewi's chemical transmitter might be involved in initiating the contraction of ordinary muscle as well as that of the heart. He set about copying Loewi's experiment using the tongue of a dog instead of the frog's heart. He chose the tongue because of its accessibility and the ease with which its nerve supply can be found and stimulated. The idea was to stimulate the nerve to the tongue and to collect the blood, as it left the tongue in the veins, in order to test it for the presence of acetylcholine. He gave the animal Dale's eserine before the experiment to preserve any acetylcholine that might be released.

The main problem facing him was how to test for acetylcholine in the presence of eserine. The only chemical method available to him could not be used in the presence of eserine. As a result he had to use a biological assay. This involved measuring the strength of contraction of the muscle belly of a leech when acetylcholine was added to the water bath in which it was suspended. This rather crude analysis worked only in the presence of fairly high concentrations of acetylcholine.

Unfortunately, in these experiments, he failed to find conclusive evidence that acetylcholine was released during the period when the nerve was stimulated. There are several possible reasons for his lack of success. The blood supply to the tongue is extravagant, as anyone who has bitten their tongue will confirm, and it is increased two- to threefold during stimulation. As a result, any acetylcholine that might have been released during the experiment would have been enormously diluted. Another problem is the varied and multiple nature of the tongue's venous drainage. These problems, together with the insensitive nature of the leech test, probably explain the failure of Feldberg's initial experiments.

Soon after Feldberg joined Dale in 1932, he persuaded him to turn his interest to the problem of chemical transmission. The two of them embarked on a further series of experiments to test their conviction that acetylcholine was involved in initiating muscle contraction. Although Henry Dale was in a pre-eminent position in the world of pharmacology and an acknowledged authority on the effect of various chemical substances on the body, his views on chemical transmission were considered iconoclastic. Apart from the group of scientists who had been influenced by Langley at Cambridge, the prevalent view in 1932–3 was that it was the electrical impulses that triggered responses in the organs, such as the muscles.

It was obvious from the beginning that Dale and his colleagues would have an uphill fight to convince the scientific establishment that nerves communicated their message by means of chemicals released at their endings. In order to persuade the doubters, they would also have to explain how these chemicals caused the electrical changes that physiologists were now able to demonstrate in both active nerves and muscle. Any evidence that they presented had to be convincing and explain the electrical changes that occurred.

The first experiments conducted by Dale and Feldberg at University College were modifications of the one that Feldberg had performed, without success, in Berlin. The tongues of both dogs and cats were used in the experiments. In order to avoid the difficulties caused by the presence of eserine, the tongues were removed from the animal immediately after it had been humanely killed. They were kept alive by perfusing them with a warm oxygenated salt solution. As the fluid collected was not blood, but a salt solution, it did not contain the enzyme choline esterase, which destroys acetylcholine. As a result, little of the acetylcholine released would be destroyed.

These experiments were a qualified success. They demonstrated, for the first time, that acetylcholine was released when the nerve to a muscle was stimulated. Unfortunately, the results were too variable to convince the doubters. In some of the experiments, a lot of acetylcholine was found after a short burst of stimulation and in others much less was produced, even though the stimulation was prolonged. The probable reason for the variable correlation between the output of acetylcholine and the amount of stimulation was that the tongue muscle tires quickly and soon stops contracting properly; as a result, the output of transmitter falls off when the stimulus is prolonged for more than a minute or so. Nevertheless, they had demonstrated that acetylcholine was released when a nerve is stimulated and the muscle it supplies is caused to contract.

The next series of experiments, which they reported in 1934, were more persuasive. In these they used the thigh muscles of a dog. The nerve to the thigh was stimulated, through a small incision in the skin, and the blood supplying the muscle was collected as it left the leg. This time they were able to show not only that the blood contained acetylcholine, but also that the amount of acetylcholine produced paralleled the duration of stimulation.

This was a major breakthrough, but it still was not enough to convince all of those who were working on the electrical changes that occurred in nerves following stimulation. The sceptics were led by an Australian Nobel Prize winner, Professor Eccles. He pointed out that it had been known for some time that acetylcholine was involved in the control of blood vessels in the muscles. He attributed the small amounts of acetylcholine that they found to an increase in blood flow produced by muscular activity. He argued that the whole process of transmitting a nervous stimulus to a muscle was far too rapid to be explained by a chemical process. He demonstrated that some muscles show a two-phase response: the first rapid, which he attributed to electrical conduction; the other slower but better sustained, which he conceded might be due to acetylcholine.

To persuade the doubters, it was necessary for Dale's team to demonstrate convincingly that the acetylcholine produced was not merely a casual accompaniment of nerve and muscle activity.

Several distinguished co-workers joined Dale's team in the 1930s. They included Vogt, Katz, Brown, Gaddum and Meledi, all of whom brought new talents and ideas to the group. He also obtained some new state-of-the-art technology that allowed the measurement of minute electrical currents by displaying them on a cathode-ray tube. With this apparatus they were able to show that there was an electrical discharge in the muscle fibres of the thigh, indicating muscle contraction, when acetylcholine was injected into the tiny blood vessel supplying it. It did not need a nervous stimulus to cause electrical muscular activity in the muscle. This experiment provided the sceptics with an explanation of the electrical activity that Sherrington had found in active muscles.

However, even showing that it was acetylcholine itself that caused muscle to contract, and not the electrical current in the nerve, did not satisfy the doubters. It was necessary to show that the acetylcholine came from the nerve when it was stimulated and was *not produced during the actual contraction of the muscle*. It became clear that the only way they could convince the doubters was to find some way to show acetylcholine was actually released from a stimulated nerve in the absence of any muscle activity.

It was at this point that curare came to play a definitive role. It is not recorded what made Dale use curare; indeed, at this time it was not certain how curare worked. It turned out to be one of those inspired leaps of faith that have played such an important part in the story of curare.

The University College team found that, when curare was given into the blood perfusing their preparation, it prevented the contraction of the muscle when the nerve was stimulated, but it did not stop the release of acetylcholine. After giving curare, they were able to demonstrate that the same amount of acetylcholine was produced when the nerve was stimulated, as was found when the muscle contracted although there was no muscle contraction. Clearly the acetylcholine could have come only from the stimulated nerve.

This evidence, together with the demonstration that when acetylcholine was injected into the vessel supplying the muscle it caused contraction, even if the nerve was not stimulated, not only proved the theory of chemical transmission but also showed the way in which curare worked. Curare had prevented the acetylcholine, released during nerve stimulation, from causing a contraction in the muscle. It had blocked the action of acetylcholine on the special sensitive area of the muscle, the muscle receptor.

Additional proof of chemical transmission and the way in which it is blocked by curare came when the same group, now led by Katz, demonstrated that the electrical response in a muscle fibre to stimulation of its nerve is progressively reduced as the concentration of curare in which it is immersed is increased. This proved that there was a quantitative relationship between curare and acetylcholine.

The messenger system

It is difficult to overestimate the importance of the finding that the brain controls the body by means of chemical transmitters. They are the brain's messengers. These messengers have become increasingly refined and added to during the evolutionary process; indeed, evolution could not have occurred without them. The evolutionary background is evident from the way in which layering of these transmitters can be demonstrated; each layer, represented by a new transmitter, indicating a new level of evolutionary development and an increase in the sophistication of the response. The bottom layer consists of the primitive acetylcholine transmitter. It is the presence of acetylcholine that causes the release of the chemical messengers adrenaline and noradrenaline, which came later in evolutionary development. Acetylcholine also controls the production of dopamine, one of the next generation of transmitter chemicals.

It is by means of these chemical transmitters that the cells of the various parts of the brain talk to one another and communicate their instructions to the rest of the body. They are the messenger systems of the body. They allow the brain to exert the central control that is necessary to keep our bodily processes working properly.

Understanding their role has given us an important insight into how the brain works and the diseases that occur when any of these systems fail.

In the years that have followed, it has been recognised that, although we are totally unaware of its actions, the brain is constantly fine-tuning the workings of the body. It uses many different chemical agents in the process. Some are slow-acting, such as the hormones; others, such as those released at nerve endings, act in milliseconds.

However, the cardinal substances, central to the whole process, is the transmitter chemical acetylcholine. The ability of curare to block the effect of this transmitter at the junction between the nerve and the muscle, the neuromuscular junction, has allowed us to study just how acetylcholine produces its effect in the body. It was this process that was finally demonstrated by Erwin Neher and Bert Sackman.

Neher and Sackman

It was the work of these two physiologists, working at the Max Planck Institute in Germany, that finally demonstrated just how the chemical transmitter worked at a molecular level and how its action was blocked by curare. They were awarded the Nobel Prize for their work in 1991. By isolating a single acetylcholine receptor from a muscle using powerful microscopes, they were able to demonstrate that acetylcholine caused the central pore, in the special receptor area on the muscle, to open, and that this permitted electrically charged sodium ions to pass through the membrane into the cell. (Fig 15) It is the passage of charged ions into the cell that causes a minute electrical impulse to be generated.

Their work finally set the seal on the controversy as to whether the brain used chemicals or electric current to transmit its instructions to the body. They demonstrated that it is the chemical transmission itself that causes the electric current that the advocates of the electrical theory had shown to occur. The electric current that they demonstrated was the effect and not the cause of neuromuscular transmission. It is an effect blocked by curare.

CHAPTER SEVEN

Serendipity

In 1929 Richard Gill was approaching middle age. He was the South American director of a successful American rubber company, which was based in Peru. He was restless; he felt his life was in a rut. His unease was made worse by a sense of a looming disaster: the early signs of the impending financial depression were unmistakable. It was this that galvanised him into making a decision that was to change his life.

During his time in South America, Gill had developed a deep affection for the countries of the Amazon basin. He especially enjoyed the lands in western Ecuador to the east of the towering, snow-capped ridge of the Andes, where the rivers that converge to form the Amazon start their descent from their origins in the nearby mountains. It was a rich, fertile land covered by forests and rolling savannah in which exotic flowers, colourful birds and animals were to be found. It was an undeveloped region with a pleasant, if rather humid, climate.

After several visits to Ecuador with his wife Ruth, he decided to give up his lucrative desk job with the rubber company and to fulfil a lifelong ambition to become a farmer, while he was still young enough to adapt to a new challenge.

He chose to build his new home in a picturesque valley, east of the Andes, where the mountains form an impressive backdrop in one direction and the rolling savannah meets the forest in the other. There he bought a 750-acre ranch. The lush land that he selected for his ranch was watered by the River Pastaza. This beautiful river starts its life in the nearby Andes and descends, watering the plains of the valley before converging to meet the Amazon. With the help of the native Indians, he built a house, the Hacienda Rio Negre, out of the wood they cut from the trees of the nearby jungle.

It did not take the Gills long to settle down to a new life on their ranch. They cleared the land and started farming bananas and coffee. Initially, they found themselves entertaining a constant steam of visitors from America, curious to see for themselves the attraction of the countryside and to hear about their new life. They mixed easily and freely with the local Indians, who were friendly and anxious to help them. They quickly learned their language and their ways. They listened to the folk tales of magic and mystery peculiar to the various tribes that inhabited that area of Ecuador.

Gill was a well-read man; he had a wide knowledge of natural history and had some basic training in medicine. He had originally intended to study medicine but had been unable to resist the lure of adventure when it was offered by the prospect of working in the virtually unknown lands of South America. He was a keen observer of the practices and rituals of the native tribes and he kept copious records of his experiences.

In his autobiography, he graphically described his many adventures and experiences during his time in the Amazon Basin. He recalls that it was not long before the local tribesmen came to accept the strangers who had made their home in their land. Because of his special knowledge of the local plants and herbal medicine and his stock of Western medicines, his services were frequently sought when sickness troubled his Indian neighbours. He became known locally as the 'white man shaman' or 'witch doctor', and he developed a reputation as a man of great wisdom. He records that, because of his status as a 'shaman', he was given an opportunity to watch the Oriente tribesmen in the secret ritual they used in the preparation of the arrow poison, which he called 'curari'.

Gill made a particular friend of a tribal leader, Servo Vargas, who was known in the local dialect as 'El Pastor'. In the summer of 1930 Vargas took him to his camp in the jungle, where he spent several days enjoying the native hospitality. It was on this trip that he witnessed the preparation of the heavy, dark-brown, gummy liquid 'curari' that the natives used to poison their arrows and darts. He was amazed at the accuracy with which they could direct their poisoned darts using primitive blowpipes made from the hollowed-out stems of jungle creepers.

He wrote that they demonstrated their skill with these darts by bringing down a bird in flight. Although they seldom missed their prey, it was not unusual for the poison to be insufficient to kill a large animal outright, but it would invariably weaken it sufficiently so that its legs gave way, making it easily caught.

After some years on his farm, he started to get homesick. The initial stream of American visitors, curious to see how the Gills had made a home for themselves in the jungle, started to dry up. He started to miss his friends and colleagues back home in America. Although he received regular letters and occasional visits from them, he found himself increasingly cut off from news of what was going on in the outside world. By 1934, some five years after he settled in Ecuador, he felt the urge to see the USA once again. He yearned for the hustle and bustle of city life and he made up his mind to visit New York.

He contacted his former office and asked them to arrange to send out a caretaker and his wife to look after the ranch while he and Ruth returned to the USA for a visit. It is evident from his writings at this time that he was anxious to explore the possibility of eventually returning to the States for good. It was while he was making preparations for his absence from his ranch that serendipity took a hand dictating a series of events that came to play a major role in the story of curare. It was these events that turned curare from a poison used by the natives to kill their prey into a drug that revolutionised medical practice.

The accident

Gill was in the habit of roaming over his vast ranch on horseback to inspect his banana plantations, castor-oil bushes and coffee crops. It was when he was returning from one such trip, shortly before his planned journey to North America, that a snake startled his horse, Chugo. It reared up, throwing him off. He landed heavily in the upright position on the heel of one foot, on top of a rough boulder. He described how at the time he felt that his spine had been given 'a quick squeeze by a giant press'. Although in considerable pain he was able to remount his horse and cover the few remaining miles to his ranch before collapsing from exhaustion.

In the days following the accident, Gill started to develop trembling and weakness in his right hand and a heaviness in his legs, which caused him to drag his feet. He started to have difficulty holding a knife and fork. A few days later he woke up to find the right side of his body paralysed. At this point he remembered his fall from the horse and the acute pain that it had caused in his spine at the time. It seemed likely to him that this was the probable cause of his symptoms.

He sought medical advice from a local doctor, who diagnosed his condition as 'spinal contusion' causing spastic paralysis. The doctor advised him to go to the USA for treatment and rehabilitation, since, in his view, the prognosis was uncertain. He returned to North America, where he spent some months lying on his back having spinal traction and remedial physiotherapy in a Washington, D.C. hospital. After some weeks he started to recover his muscle power, but full recovery from the paralysis was slow.

It was some three months before he was able walk unaided and almost four years before he was fit enough to consider returning home. He never completely recovered from the accident. In later life he had episodes of weakness and trembling in his hands and legs. In retrospect, it is possible that it was not the fall from his horse that caused Gill's problems but that he was actually suffering from a form of multiple sclerosis. During his stay in hospital Gill used his time to write a series of delightful children's stories based on the native legends and superstitions that he had heard told when he had been a guest of his Indian friends in the jungle.

It was during this recovery period in the USA that Gill suffered severe agonising muscle spasms in his right arm and leg. The painful spasms that he experienced are not uncommon in patients developing or recovering from spinal injury. They are not unlike the painful episodes experienced by healthy individuals at night in their calf muscles when a slight movement triggers off a cramping pain without any warning. Although these calf cramps are agonising, they are mercifully of short duration; even so they leave the muscles aching for many minutes after the pain has dispersed.

However, the spasms associated with injuries to the spine often cause a pain that lasts for hours at a time. It was because of these distressing attacks that Gill's thoughts turned to the possibility of using the muscle-paralysing action of curare to reduce the severity of the spasms. He knew that curare had been used by Dr Michael Burman in America a few years earlier to control muscle spasms in children with spastic disorders with some success, and he had observed, at first hand, curare's paralysing effect on animals during his days in the jungle with Servo Vargos.

Gill's inspired idea of using a paralytic agent to reduce the spasms was many years ahead of its time. In recent years, in order to avoid the risk of causing a widespread paralysis by drugs like curare, botulinus toxin, whose effect is localised, has been used. The toxin acts locally on the muscles in a manner similar to curare. It is widely used in pain clinics to reduce painful muscle cramps. Botulinus toxin produces a similar paralytic effect to curare, although it works at a different point in the various stages involved in the transmission of nerve signals to muscles. It is this that has led to the use of botulinus toxin, under the trade name of Botox, for cosmetic purposes, to paralyse those muscles of the face that are causing wrinkles.

The expedition

It was the prospect of using curare to reduce the painful muscle spasms that made Gill determined to return to

Ecuador to collect sufficient curare to try out its use in patients. He started to prepare for the adventure while he was still in hospital. It took him some time to mobilise the finance necessary, but, by 1938, he was ready and able to mount an expedition to collect as much curare as possible so that it could be purified and undergo clinical trials in the States.

In this endeavour he was backed by the Boston philanthropist Sayre Merrill, who promised to finance the expedition provided he was able to obtain the backing of a pharmaceutical company that was prepared to handle the curare he brought back. He negotiated backing from Merck & Co. with a promise that they would test the effect in their laboratories. At the time Merck had initiated a programme for investigating the pharmaceutical properties of native herbal remedies to find out if they had any activity that could be exploited commercially.

The expedition that Gill put together in 1938 consisted of more than ninety tribesmen and nine large dugout canoes. The canoes were loaded with medicines, sewing materials and the gifts that they intended exchanging for curare resin. They set off on their journey to the headwaters of the Amazon in late autumn 1938. It took them five months to complete their mission. They succeeded in bartering their goods for about 30 pounds (some 13 kilos) of crude curare, which they loaded onto the canoes for the return journey.

While they were collecting the curare Gill collected specimens of the various creepers from which the crude curare had been made. He did this at insistence of the Merck company's botanist, Dr Boris Krukoff, who wanted to identify, once and for all, the actual source of the poison. Gill readily agreed to his request, as he was well aware of how important it was to overcome the uncertainty as to which creeper was the true source of the paralysing resin and which was the cause of the convulsive impurities found in some of the crude native preparations.

It is likely that, had Gill's horse not stumbled and thrown its rider in a remote part of Ecuador in 1934, the history of modern medicine would not have developed in the way it has. At the time, curare was little more than a pharmacological curiosity. Although there had been attempts to utilise its paralytic properties in medicine, they had met with limited success. It was serendipity that produced the convergence of events that led to a large quantity of the drug arriving in the United States in 1938. Again, it was by good fortune that it eventually ended up in one of the few laboratories where the pharmacologists had sufficient insight to envisage its potential usefulness in medicine.

Purifying the curare

When Gill returned to the USA in 1939, he found that the Merck company were no longer interested in curare. They had turned their interest to different plant, whose extract had a similar paralysing property. Gill was naturally very upset by their lack of interest but, undeterred, he immediately started sounding out other companies in an effort to find someone ready to purify and test his curare.

After months of negotiation he managed to persuade E. R. Squibb & Son to purchase his curare with the promise to try it out as a cure for painful muscle spasms. As a result it was in the laboratories of E. R. Squibb, in America, that curare was first purified and standardised with a view to producing a commercial preparation that could be used in clinical trials. Meanwhile, the plant extract that Merck & Co. had been investigating, erythroidine, was abandoned as not being commercially viable.

The first task was to separate the pure paralysing curare from the other chemicals in the crude resin. Gill was well aware of the danger of using the impure, unstandardised drug. He wrote in his delightful book *White Water, Black Magic*:

> *Lack of uniformity in the curari samples available to research has proved a great obstacle in its clinical use. Many kinds of foreign matter are sometimes added to the alkaloid during its manufacture for reasons of carelessness or superstition.*

Isolating the active ingredient in the resin was not easy. It was not the first time an effort had been made to separate it from the mass of impurities. An attempt had previously been made by Dr Boussingault, working in the South American city of Bogotá in 1827. He produced an extract that he called 'curarina' from specimens of 'woorari gum'. Although it was free of convulsive contaminants it was unlikely to have been pure curare. His efforts were said to have been made at the instigation of Simón Bolívar, the guerrilla leader who became the founder of Bolivia. Samples of his purified drug reached England and are referred to by Thomas Spencer Wells in 1859 in one of his reports to the Royal Medical and Chirurgical Society of London.

In 1886, Dr Rudolph Boehm took the process one step further by attempting to separate the effects of the active ingredient from the common contaminants found in the gum. He found that the contents of the crude poison varied among the different samples of resin. He believed this was due to mixing the extracts from several different plants. The plants used varied according to the area where the poison was made. He thought, erroneously, that this was reflected in the nature of the container used to store the poison.

He labelled some of his specimens as *tube curare*, some as *pot curare* and others as *calabash curare*. It was Gill who pointed out, some years later, that Boehm was wrong in his assumption. He found that the Indians he watched used whatever container was near at hand to store their poison and that the container did not necessarily identify the nature of its content or the area from which it was collected. Boehm did succeed in preparing crystalline 'curarine' and analysing its chemical content. However, it was not, as he believed, pure. It contained significant traces of other salts. He failed in his attempt to establish the chemical nature of the drug, although he confirmed the fact that it was an alkaloid.

The chemistry of curare

It was due to the work of Harold King, working at the National Institute for Medical Research and the Burroughs Wellcome laboratories in London in 1935, that the structure of the chemical was recognised. King worked with an old specimen of curare that had been made from the creeper *Chondrodendron*, which he had found in the Burroughs Wellcome museum, where it had the enigmatic label 'Ucayalli river 1871'. Because it had been stored in a hollow bamboo tube, the purified alkaloid was called *tubo-curarine*.

It was found that as little as 10 mg of pure tubo-curarine would paralyse a dog. King's formula showed that the chemical was like two molecules of acetylcholine, a simple natural chemical produced by the action of acetic acid on ammonium, lined up and separated from each other by a bulky but inert chemical scaffold.

Once its structure was demonstrated, it became a template for the production of other compounds of a similar conformation by medicinal chemists. It was soon realised that the essential ingredient for a successful synthetic curare was the presence of two acetylcholine-like molecules separated from each other by a fixed, pre-determined distance. The two acetylcholine-like molecules can be looked on as two hooks, and the chemical structure in between as a stick that has to be of a certain length so as to separate the hooks from each other by a set distance. This allows the hooks to latch onto two specific pegs at the receptor simultaneously. When both hooks are engaged and fitted onto the pegs, a neuromuscular block ensues. If the stick is either too long or too short, only one hook will be able to engage with a peg at a time and paralysis will not be produced.

Although it is now known that the King formula is not entirely accurate, it had the effect of initiating the search for synthetic alternatives to curare that were cheaper and more selective in their actions. It was the start of the search for drugs that had a specific action at one site only and were free of unwanted side effects – so-called designer drugs.

CHAPTER EIGHT

Curare in Anaesthesia

Intocostrin

The first reported use of Richard Gill's curare, purified by Squibb and presented as a standardised drug called Intocostrin, came from the Orthopaedic Hospital of the Nebraska State University.

In 1938, the professor of psychiatry at the university, Dr A. E. Bennett, used an infusion of dilute intocostrin to reduce the convulsions associated with Metrazol shock therapy. Metrazol (metranidazol) is a strychnine-like convulsive agent, which was used to produce short-lived convulsive seizures as an alternative to electroconvulsive therapy (ECT). It was preferred by some to ECT, as it was claimed that it produced less loss of memory. Although the treatment was controversial, it often produced a dramatic, if short-term, improvement in patients with schizophrenic disorders and depression, especially in those at risk of committing suicide.

Bennett had heard about Gill's curare from his friend,

the neurologist Dr Walter Freeman, whom Gill had consulted about his neurological problems when he returned to the United States. Bennett described the technique he used to administer the intocostrin in great detail. He infused it intravenously, diluted in an alcoholic solution, until the patient could no longer lift his head from a pillow for more than a minute or two.

At this point he stopped the infusion and gave the Metrazol. Not only were the subsequent convulsions much reduced but, somewhat surprisingly, he found the convulsent drug appeared to help the patients recover their muscle power more rapidly. He estimated that the dose of intocostrin he used was less than half that which would have stopped his patients from breathing. Before the advent of curare, the muscle spasms that occurred during the seizures induced by Metrazol shock therapy were so severe that about a quarter of the patients suffered compression fractures of their vertebrae. The patients had to be strapped to the bed to prevent the fits it induced from throwing them on to the floor.

Bennett's use of curare was based on the reports in the literature of the work of Dr Michael Burman, who had used a tiny dose of a crude preparation of the poison[4] in several children with spastic muscle disorders some years earlier. He reported that it reduced their muscle spasms so that they could move their limbs more freely and that, in the dose he used, it was without apparent ill effect.

4 It is probable that Burman used the drug erythroidine, which Merck & Co. were purifying from another South American plant.

In the first paper on the subject published by Bennett, not a single patient in his series who underwent Metrazol convulsive therapy suffered a fracture in spite of using a dose of curare that he estimated was less than half the dose that caused paralysis.

Why was Gill's curare first used in Nebraska? The answer lies in the presence in that hospital of a professor of pharmacology and physiology, Dr A. R. McIntyre, who, working with a research grant from Squibb, was studying the chemistry of nerves and the chemicals released when they were stimulated. He was aware of the pioneering work of Henry Dale and his team in London, who had established that the chemical acetylcholine, a simple chemical made from the combination of an ammonium radicle and acetic acid or vinegar, was produced when a nerve to a muscle was stimulated. Dale had proposed that it was a 'chemical messenger' that carried the message from the nerve to the muscle.

It was Dr McIntyre who was consulted about the purification of Gill's curare in the Squibb laboratories. He had co-operated with Dr Horace Holaday in developing a standardisation procedure that was based on the amount of drug that prevented a rabbit from holding up its head – the so-called 'rabbit-head-drop test'. Doubtless it was due to his presence at the university and to Bennett's concern about the fractures in his patients that curare was first used, under clinical conditions, in Nebraska.

Lewis Wright

The introduction of curare into anaesthesia owes much to the enthusiasm and persistence of Dr Lewis Wright, a doctor working for E. R. Squibb & Son, whose contribution to the curare story has not been sufficiently well recognised.

Wright had spent some time training as an anaesthetist at New York's Bellevue Hospital, before going on to complete his residency programme in obstetrics. At the time of World War Two, the hospital was one of the few teaching institutions in America to have a recognised anaesthesiology residency training programme. The war found the USA desperately short of medically trained anaesthesiologists. In the USA, nurse technicians had administered almost all the anaesthesia for surgery before the war. They were ill prepared for the requirements of surgery in badly injured and shocked servicemen. The high number of anaesthetic deaths among wounded servicemen, following the attack on Pearl Harbor, became a scandal that exposed these deficiencies.

Towards the end of the war, training centres were established by specialist, medically trained anaesthetists, such as Dr Emanuel Papper in New York, Dr Henry Beecher in Boston and Dr Ralph Waters in Wisconsin. Many of these founding fathers of modern American anaesthesia had received their training in the armed services. They initiated the training programmes that were to establish the reputation of American anaesthesia in the years following the end of the war.

There is little doubt that Wright had been impressed by the scientific rigour of his teachers during his anaesthesiology residency at the Bellevue Hospital in 1939. He had first-hand experience of some of the problems associated with producing good conditions for surgery. He saw the possibility of using curare to produce paralysis of the stomach muscles for intra-abdominal surgery in the way that Bennett used it to prevent convulsions in shock therapy.

At this time deep levels of anaesthesia were used to obtain sufficient muscle relaxation to allow the surgeon to work on the bowel, and especially to return distended loops of intestine into the abdominal cavity at the completion of an operation. Deepening the anaesthetic at the end of a procedure to allow the surgeon to sew up the abdominal muscles was fraught with danger: it invariably weakened the patient and delayed his or her recovery from the effects of the anaesthetic.

Initial trials in anaesthesia

Lewis Wright had observed at first hand the successful use of curare by Bennett. He had been impressed by the muscle relaxation it produced and the safety of its administration in the way Bennett used it. Because of his anaesthetics background, he immediately saw that it might be used to produce muscle relaxation during surgery with the need to give very deep anaesthesia to patients who were sick. He set about persuading some of the leading figures in American anaesthesia to try the drug on patients.

He attended the American Society of Anesthesiologists meeting in New York in 1940, on behalf of the Squibb company. Here he showed a film of Dr Bennett using intocostrin in Metrazol shock therapy. At the meeting he met his old chief, Dr Rovenstein, from the Department of Anesthesiology at Bellevue Hospital. Rovenstein immediately grasped the potential advantages of a drug that produced muscle relaxation without the necessity for deep anaesthesia. He determined to try it out in patients.

Armed with samples of intocostrin provided by Wright, Rovenstein and his assistant, Dr Emanuel Papper, first tried out its effect on dogs. The results of the experiments were disappointing. To their amazement, as soon as the dogs were injected with intocostrin they developed an attack of wheezing, asthmatic breathing, which they diagnosed as 'bronchospasm'. In spite of this setback, they decided to try it on patients.

Some weeks later, Rovenstein gave 5ml of the curare preparation (about the same dose as used by Bennett) to a patient anaesthetised with ether while Dr Papper looked on and kept a record of the patient's pulse and respiration. With increasing alarm they saw the patient's breathing become shallower and shallower until finally it stopped completely. The operation was abandoned in panic and the patient given artificial respiration until he slowly started to breathe again on his own.

Undeterred by this near disaster they tried intocostrin, in a smaller dose, on another patient with a similar, if less alarming, result. Rovenstein was so disturbed by these events that he never repeated the experiment. He explained that he felt the margin of safety of the drug was too small for it to be used in sick patients. He attributed the profound effect of the small dose of drug that he used to the relative weakness of his patients compared with those receiving it for Metrazol therapy.

These early pioneers were unaware of the considerable additive effect of ether and curare. Indeed, it was not until a year or so later that the potentiating effect of ether on curare was demonstrated. What was too small a dose of curare to cause paralysis in Bennett's unanaesthetised patients caused profound paralysis when given to Rovenstein's patients, because they were under ether anaesthesia.

In order to understand the anxiety of these two very capable anaesthetists it is necessary to appreciate the difficulties facing them. At the time the shallow breathing and respiratory failure they observed was usually associated with too much anaesthesia. It was often the harbinger of death in sick patients. In present-day anaesthetic practice most patients have a breathing tube inserted into the trachea in a procedure known as *endotracheal intubation*, which makes it a simple matter to support shallow breathing and respiratory failure by artificial ventilation. This type of anaesthesia, using the trachea tube, was rarely practised in America at the time.

As a result of his frightening experience, Rovenstein suggested that, if drugs of this type that affected only the abdominal muscles and not the respiratory ones could be developed, they would be extremely useful in anaesthesia. This was a view that was to muddy the waters and influence much of the research into safer substitutes for curare for many years as researchers searched for muscle relaxants that had these selective properties.

At the same meeting in New York, Lewis also met another leading figure in American anaesthesia, Dr Stuart Cullen. Cullen was one of the few anaesthetists in America at that time who were well versed in research techniques. The school at Iowa, where he trained and worked before becoming the chief of anaesthesiology in San Francisco, was one of the leading research centres in the States, with a pharmacology department that had a formidable reputation. Cullen was therefore well placed to try the effect of the sample of intocostrin that Wright gave him.

He tried out the effect of the drug on dogs. As soon as the drug was injected, he noticed a profound and prolonged fall in the animal's blood pressure. He also noticed the bronchospasm that Papper had seen. As a result, after several experiments, he came to same conclusion as Rovenstein: that the drug was too dangerous for use in sick patients during anaesthesia.

He was obviously influenced by Rovenstein's reports. Like him, he argued that Dr Bennett's patients had not been physically sick and were therefore better able to withstand the deleterious side effects of the drug. Neither Cullen, nor Rovenstein nor Papper, was aware that the effects they found on the breathing pattern of dogs are the result of their peculiar sensitivity to a side effect of curare that is rarely seen in man. In many animals curare causes a release of histamine, producing an effect not dissimilar to an acute asthmatic attack, and it causes a fall in blood pressure. In spite of these setbacks when he tried the drug experimentally, Cullen was one of the first American anaesthetists to embrace the use of curare, once its side effects had been tamed and the ill effects he had seen were explained.

The breakthrough

Lewis Wright was unperturbed by these rebuffs. He had witnessed its safe use by A. E. Bennett in Nebraska and was convinced of its potential importance. He continued to seek out anaesthesiologists at the meetings of the American Society of Anesthesiologists meetings and to try to persuade them to try out intocostrin on their patients. However, news of the difficulties experienced by Rovenstein and Cullen, two of the most respected figures in American anaesthesia, invariably deterred them from trying the drug themselves.

At one of these meetings he met an old friend, Dr Harold Randall Griffith, who had recently become chairman of the Department of Anaesthesia at the Homeopathic Hospital in Montreal (later renamed the Queen Elizabeth Hospital). Griffith was a cautious clinical anaesthetist trained in the British pattern of physician anaesthesia. He had been responsible for introducing the newest anaesthetic agent, cyclopropane, into Canadian anaesthesia. He was used to keeping detailed records of his anaesthetics and, unlike many of his American contemporaries, he was well used to intubating the trachea of his patients and was therefore well able to deal with any respiratory failure by assisting the patient's breathing.

Before he tried Lewis's new drug, he made careful enquiries about its effects. At this time Dr McIntyre was working in Montreal, in the University of McGill Department of Pharmacology, with Dr R. J. Birks, a distinguished neuropharmacologist. They were interested in the mechanism underlying the release of acetylcholine from the nerve endings and the effect of curare on this process.

It is tempting to speculate that their presence may have influenced Griffith's decision to try out curare. He did so with some misgivings and only after much preparation in case anything should go wrong. He used the drug on a fit twenty-year-old plumber undergoing an appendectomy at the Homeopathic Hospital on 23 January 1942. It is important to note that, unlike Rovenstein, who used ether anaesthetic, he anaesthetised his patient with cyclopropane, an agent that does not potentiate the paralytic effect of curare. He records:

> *Intocostrin [the Squib preparation of curare] 3.5 ml given intravenously – no appreciable effect on pulse or respiration. After 5 mins a further 1.5 ml of intocostrin given. Apparent complete relaxation of the abdominal muscles resulted and continued for 20 min during which*

time the cyclopropane anaesthetic was lightened. At the end of this period muscle tone returned…

Together with Dr Enid Johnson, he went on to use intocostrin in twenty-five patients without any complications. They did not see the profound depression of respiration reported by Rovenstein. Their results were published in the *American Journal of Anesthesiology* in 1942. The publication of their paper opened a new era for anaesthesia. It changed for ever the way it is administered. It was justly hailed, by two famous British anaesthetists, Cecil Gray and John Halton, writing in the *Journal of the Royal Society of Medicine*, as 'a milestone in anaesthesia'. A new chapter had been opened in the practice of anaesthesia.

CHAPTER NINE

A Milestone in Anaesthesia

The accolade for the introduction of curare into anaesthesia rightfully belongs to Harold Griffith and Enid Johnson of Montreal. However, had it not been for the foresight and persistence of Dr Lewis Wright of E. R. Squibb & Son, it is possible that no one in America would have been prepared to use the drug after the scares that had accompanied its initial trials. It took a committed enthusiast to persuade yet another anaesthetist to try curare after two of the leaders of American anaesthesia, Dr Rovenstein in New York and Dr Cullen in San Francisco, had tried it and pronounced it too dangerous to use in sick patients.

One of the factors that complicated the scene at the time was the war in Europe. This made it virtually impossible to try out the drug in the United Kingdom. Britain would have been a natural test site for a new type of anaesthetic drug, even for an American company. Since the days of John Snow, Britain was the only country where anaesthesia was almost exclusively administered by qualified doctors and where a body of professional, medically trained anaesthetists, with a burgeoning reputation for scientific research, could be found. This was in spite of there being no legal requirement for an anaesthetist to be medically qualified, since all that was necessary before administering anaesthesia was for an unqualified practitioner to tell the patient he was not a doctor. As a result, the standards of clinical practice were the best in the world and the anaesthetists among the foremost pioneers of new drugs and techniques.

Unfortunately, the war prevented the limited available medical resources from being diverted away from treating the sick and wounded in order to test new drugs. Wright shrewdly realised that Canadian anaesthetists ranked highest in the league of likely pioneers and they were close at hand. In Canada, anaesthetists were, for the most part, medically trained professionals. They were more influenced by the developments in the UK than the USA and followed the British methods and practices.

Harold Griffith was well placed to try out Intocostrin, the Squibb preparation of curare. He was a respected, cautious anaesthetist who enjoyed a close friendship with many British colleagues. His position as chief at the Montreal Homeopathic Hospital was politically secure. His brother was the principal surgeon at the hospital and his father had been its first medical director. There is no doubt that Griffith knew of both Rovenstein's and

Cullen's work and their misgivings, but he was persuaded of its overall safety after seeing the film of its use by Bennett to control the convulsions caused by Metrazol in psychiatric patients.

Griffith had been responsible for the introduction of the new gaseous anaesthetic agent cyclopropane into Canada. He had first seen it demonstrated by Ralph Waters in Wisconsin, USA, and appreciated the flexibility and the ease with which it could be used to induce anaesthesia. It had largely replaced ether in his practice. As a result when he came to try out Intocostrin, unlike Rovenstein – who used ether as the anaesthetic agent – he used cyclopropane as the basis for his anaesthetic technique. This proved to be a piece of good fortune, as cyclopropane, unlike ether, does not potentiate the paralysing effect of curare.

When he had first used Intocostrin to relax the abdominal muscles of the plumber in 1942, he had a laryngoscope ready should the patient's respiration become depressed and should there be a need to intubate the patient's trachea to administer artificial ventilation. The Intocostrin worked well and he was able to produce good surgical conditions with its help, at much lower concentrations of cyclopropane than usual. He and Johnson went on to use the Squib curare preparation in twenty-five patients before reporting their results in July 1942 in the American journal, *Anesthesiology*. In that paper they report, somewhat modestly, 'We have been so impressed by the dramatic effect produced in every one of our patients that we believe this investigation should be continued.'

In terms of present-day practice in research it is interesting to reflect that no permission was sought for the use of this new and potentially dangerous drug, either from the hospital or the patients involved. In 1942, there were no ethics committees from whom approval had to be obtained for the trial of a new drug. It is extremely likely that, had the situation arisen today, they would have been prevented from pursuing the trial without first carrying out animal work to make sure it was safe in all conditions, in all patients. Like other researchers, they would have been likely to have tried out its effect on dogs.

Both Rovenstein and Cullen had demonstrated that curare causes serious side effects in these animals due to the release of histamine. Under these circumstances it is highly probable that the world would have been denied the benefits of curare for several years.

Much to his credit, on reading the report of the successful use of curare by the two Canadian anaesthetists, Professor Cullen took up the challenge and tried the drug once again, this time on patients in San Francisco. He published his experience with its use in 131 patients in 1943. His change of heart on its safety gave the seal of approval to the new agent. Two years later he reported its use in a thousand cases. The report signalled the go-ahead for the widespread introduction of curare in the USA.

Almost overnight it became standard practice to give

intocostrin to patients for all abdominal operations. It was taken up both by experienced physician anaesthetists and by less well-trained nurses, who often failed to appreciate the potential risks associated with its use.

It was not long before the potential benefits of the new drug were being exploited in England. In 1946, one year after Cullen's report, Professor Cecil Gray and Dr Jack Halton reported the use of a newer, purer curare – Burroughs Wellcome's d-tubocurarine chloride – in a thousand cases in Liverpool. Their paper – presented to the Royal Society of Medicine in London and entitled 'A Milestone in Anaesthesia?' – set the scene for its widespread use in Europe.

Curare use in Europe and America

Although the British anaesthetists were encouraged to try curare by the reports from the USA, they were not influenced by the fears that had dogged its early experimental use in America. Very soon it became evident that there was a very different approach to the use of curare on the two sides of the Atlantic.

Anaesthetists in America looked on curare as an inherently dangerous drug and used it with extreme caution. This was probably a result of the early experiences of Rovenstein and Cullen. At the time most American anaesthesia was administered by special nurses, who lacked experience at intubating the trachea and administering artificial ventilation of patients with respiratory depression. They were taught that any reduction in the depth or rate of ventilation was to be strictly avoided. In their hands artificial ventilation was highly dangerous. They recommended that only a very small dose of intocostrin should be used, one that fell well short of the amount necessary to paralyse the patient.

It was taught that total paralysis should be avoided at all cost; it was a step into the unknown. It was emphasised that the muscle relaxant was to be used to *supplement* the effects of an anaesthetic agent, not to replace it. As a result the same deep levels of anaesthesia were used, and the anaesthetic was invariably deepened at the end of an operation in order to obtain extra muscle relaxation so that the distended bowel could be returned to the abdominal cavity. As a result, they missed out on much of the benefit offered by the new drug.

In America, it was taught that if the patient's breathing became too shallow it should be gently assisted by squeezing the anaesthetic reservoir bag. The importance of 'keeping in touch' with the patient's own respiratory pattern by means of what was termed *assisted ventilation* by 'the educated hand' became part of the dogma associated with its use in the USA. However, in practice, as most nurses used a face mask to administer anaesthesia to their patients, it was often difficult to hold the mask on the patient's face while squeezing the reservoir bag in time with the patient's breathing.

In the United Kingdom the approach was quite different. The teaching in Britain was that a drug should be used in a dose that produced optimum results. A paralytic drug should therefore be used in a dose that produced paralysis and the respiratory paralysis that it caused should be dealt with by artificial ventilation. As Gray and Halton pointed out in their seminal paper, with this technique it was possible to use very much lighter planes of anaesthesia. This meant less physiological disturbance of the body and a more rapid and comfortable recovery.

This was the technique that found favour in most centres in Europe. The dose of curare used in Europe was often five to ten times that used in America. The European approach brought better relaxation without the need for the deep planes of anaesthesia used in America. The disadvantage of the technique was that the effects of the large dose of curare had seldom completely worn off by the time the operation ended. Unless the residual effects of the drug were reversed satisfactorily at this time, the patient was left breathing inadequately, unable to cough up sputum and in a position where he was vulnerable to inhaling vomit.

Because of the higher doses of curare used and the likelihood of some residual paralysis at the end of the operation, it was frequently necessary to give a curare antidote, an anticholinesterase drug such as neostigmine. These drugs were developed from eserine, the drug that Henry Dale demonstrated blocked the action of the enzyme that destroys acetylcholine. Unfortunately, there was a learning curve to be overcome for the safe use of neostigmine, and it was some time before the dangers that could accompany its use in patients were fully recognised. During this time its use was discouraged in the USA.

The American technique also brought problems. It was not always easy to assist the ventilation of patients who were not fully relaxed. This frequently led to underventilation and poor oxygenation of the patient. This was especially liable to occur when nurse anaesthetists, who were inexperienced in the technique of assisted ventilation, used curare.

In America the failure to use antidotes often meant that patients were left partially paralysed and unable to cough properly or fend for themselves at the end of an operation, especially if the procedure had been short.

Present-day practice

Over the years since the introduction of curare, there has been a coming together of the different methods of using the drug. Today, new muscle relaxants, the successors to curare, are always administered in a paralytic dose and the anaesthetist always takes over the control of the patient's breathing. The very large doses advocated by Gray have become moderated and ventilation is now carefully adjusted to maintain a normal level of carbon dioxide in the blood so as to avoid a disturbance of the body's chemistry.

The vast majority of patients now receive a muscle relaxant to paralyse them at some time during their anaesthetic and an anticholinesterase drug at the end of the surgery. The side effects and complications of their use are so low that they are one of the safest drugs used by anaesthetists.

The control of ventilation

In the 1950s there was a sudden increase in the number of cases of poliomyelitis in the Western world. A particularly bad epidemic occurred in 1952 in Denmark. So many patients developed respiratory paralysis, requiring artificial ventilation, that the limited supply of iron lungs was soon used up. At the time infectious diseases such as polio were treated in isolation units separate from the mainstream of the hospital.

As a result of this potentially catastrophic shortage of iron lungs, help was sought from outside the infectious-disease unit. Other disciplines in the hospital were asked to help. It was eventually brought to the attention of the anaesthetists who, thanks to their experience with curarised patients, were used to treating paralysed patients. They recognised that the ventilatory paralysis of the polio patients could be treated in exactly the same way as it was in curarised patients.

An emergency team was set up and the polio patients were ventilated, using an anaesthetic machine connected to a source of compressed air, by relays of doctors, nurses and students who took it in turns to pump air into their lung. It had taken about 140 years for us to rediscover that the best way of treating the victims respiratory paralysis is by pumping air into their lungs in the same way as Benjamin Brodie had in 1812 when he used a domestic bellows to ventilate a 'she ass' he had curarised. The only difference was that the Danish anaesthetists squeezed the gas reservoir bag on an anaesthetic machine circuit as a means of inflating the lungs rather than a bellows.

Following the experience in Denmark, it did not take long for the iron lung to be condemned to history. Mechanical devices were developed to replace manual ventilation and to pump air, or anaesthetic gases, into the patient's lungs under 'positive pressure'. Some of the early ventilators were very simple and unsophisticated. One of the first of these machines, developed in London by Dr Atwood Beaver, used the windscreen-wiper motor from an Austin Seven to drive the pump, while another used the gearshift from a bicycle to change the ventilation rate! Today the machines are much more complex and have the ability not only to alter the rate and the volume of gas pumped, but also the way in which the inspiratory and expiratory phases of ventilation are timed so as not to embarrass the circulation by producing too much pressure in the chest.

Artificial ventilation

It is a result of the knowledge gained, following the introduction of muscle relaxant drugs into anaesthesia, that artificial ventilation can now be safely maintained for days or months, using positive pressure. Today, patients paralysed by muscle relaxants can undergo the most difficult and dangerous surgery under light levels of anaesthesia. As a result, there is far less disturbance of their metabolism and their physiological parameters remain within normal boundaries, even during the most prolonged and difficult surgery. They recover rapidly from the effects of the anaesthetic and in most cases they wake up clear-headed, without nausea and vomiting. Whereas in the past only relatively healthy patients were considered 'fit for anaesthesia', today there is no such state as 'unfit for anaesthesia'.

Just over twenty years after the introduction of curare into anaesthetic practice in the UK, a study reported to the Royal College of Surgeons in 1971 revealed that anaesthetic mortality had been reduced by almost 30 per cent. Most of this reduction can be attributed to the benefits of the muscle-relaxant drugs and our better understanding of human physiology that resulted from their use. Truly, the introduction of curare into medicine had produced a 'milestone in anaesthesia'.

CHAPTER TEN

Problems

Complications

Whenever a new, effective drug is introduced into medicine it is followed by a wave of excessive enthusiasm. As a result, it is often used in inappropriate circumstances or in improper doses. This then produces reports of complications, which get prominence in the medical press, causing a decline in the use of the drug. This is followed by a period in which common ground is agreed on the correct way the drug should be used, the doses in which it should be administered and the contraindications to its use.

It is only after the limitations of the drug are appreciated, and its use restricted to those conditions where it is safe and effective, that one can truly assess the role of the new agent in the treatment of patients. So it was with curare.

The Beecher and Todd report

The first influential report into the side effects of the use of curare in America came with the Beecher and Todd report in 1954. This report was published as a paper in the *Annals of Surgery*. Since it came from two distinguished Harvard doctors, Henry Beecher and D. P. Todd, it received considerable publicity.

The paper studied the death rate that followed the introduction of curare in ten university teaching hospitals in America. It covered the period from 1948 to 1952, during which time just under a total of 600,000 anaesthetics were given in these institutions. They found that, since the introduction of curare into anaesthetic practice, the death rate from surgery had risen sixfold.

While no one challenged these findings, the assertion that it was the curare that had somehow poisoned the patients was contentious. It was pointed out that only about 10 per cent of the anaesthetics reported were administered by trained anaesthetic physicians at the time. Most anaesthetics were given by nurses who had not been trained to administer artificial ventilation or to recognise the signs of underventilation. Too often the patient would be anaesthetised, partially paralysed and left hardly breathing during the surgery either because the reparatory depression was not recognised or because it proved too difficult for the nurse to assist the patient's spontaneous breathing using a face mask.

In contrast, in England, where the patients were given a large dose of curare and it had became a habit to reverse any residual blockade at the end of the operation, the death rate after surgery, during the years following the introduction of curare, actually fell.

In America, the teaching was to avoid the use of anticholinesterase drugs whenever possible, as they were considered an additional hazard. This resulted in many patients being left partially paralysed during the recovery period. In retrospect, there is little doubt that Beecher and Todd's findings were the result of the poor standard of anaesthetic care in America at the time, due to the lack of professional, well-trained physicians.

Beecher and Todd's findings, coupled with the fear of litigation, caused the use of curare to decline rapidly in the USA after 1954. The experience in Britain was quite different. As its benefits became apparent, curare soon became the basis of most anaesthetic techniques. However, even in Britain it was not without its problems.

Neostigmine-resistant curarisation

In 1956, Andrew Hunter, a distinguished British anaesthetist, described a syndrome he termed *neostigmine-resistant curarisation* in six patients, all of whom subsequently died. The syndrome followed a characteristic pattern. All the patients were elderly or seriously ill; they all had suffered from two to four days of diarrhoea and vomiting caused by an obstruction of the bowel. The operation to relieve the problem was carried out without difficulty and the anaesthetic technique used curare in all except one case, where a new artificial curare-like drug was given.

The procedure was carried out without any noticeable complications until the time came to reverse the paralysis.

In all the cases described by Hunter the patients started breathing following the administration of the anticholinesterase, neostigmine. In every case the breathing was shallow and laboured. Every breath was associated with a downward tug on the jaw and trachea. It was as if the whole content of the chest was jerked down every time the diaphragm contracted. Further neostigmine failed to make the situation any better. After a period, ranging from minutes to hours, the heart started to beat irregularly and the blood pressure fell. All the patients eventually died of cardiac failure.

Within six months, forty-one further cases, all conforming to the same general pattern, had been reported in the medical journals. Although most of these cases came from Britain, some came from continental Europe and South Africa. Puzzling though these cases were, they might have been dismissed as part of the learning curve for curare had it not been that similar cases were not seen in the USA.

Various suggestions were made as to the cause of the syndrome. Attention focused on the use of the large doses of curare in Europe and the suggestion was made that the syndrome represented an inherent toxicity of the drug that became apparent in sick patients, possibly as a result of the curare acting on the brain when these large doses were used. It was known that in normal circumstances curare did not penetrate the covering layers of the brain.

When Intocostrin was first introduced into anaesthesia, attempts had been made to see just how safe it was. In

1950, F. Cole gave up to twenty times the paralysing dose of Intocostrin to dogs that were artificially ventilated. Although he found it caused breathing difficulties and bloody diarrhoea, associated with histamine release, he found that even in these huge doses it did not kill the animals. Many years later, when I worked with Ellis Cohen in America, we demonstrated that, even in large doses, radioactive curare does not penetrate the lining membranes of the brain.

Research into neostigmine-resistant curarisation

In 1957 I was fortunate in obtaining a fellowship to study research at the University of Washington, Seattle, under its very able director, Lucien E. Morris. He taught the American way of using curare while I had been schooled in the European technique. In light of Hunter's neostigmine-resistant curarisation syndrome, I realised that one could justify the use of high doses of curare only if it could be shown that it had no inherent toxicity, even when given in very large doses.

In view of the suggestion that these large doses might affect the brain, we centred our study on the central-nervous-system effects of the drug. Under the careful scrutiny of Professor Evan Frederickson, I set up a series of experiments to study the toxicity of curare under carefully controlled conditions. In many ways our experiments simulated, in a more up-to-date and sophisticated manner, those of Benjamin Brodie a hundred years earlier.

We used cats in these experiments as, unlike dogs, they do not respond so violently to curare by releasing histamine, and their sensitivity to it is similar to that of man. After very few experiments we demonstrated that, even if fifty times the paralytic dose of curare was given, provided the animal was given fluid and the ventilation was carefully controlled over the twenty-four to thirty-six hours that they were paralysed, normal recovery was possible (under the terms of our licence we could not allow complete full recovery to take place). During this time the circulation functioned normally.

Although changes did occur in the electrical activity of the brain, they suggested a deepening of the anaesthetic state rather than a toxic effect. They were always reversed when the effect of the curare started to wear off. These observations convinced us that it was not an inherent toxicity that had caused the problems Hunter had described. If it was not due to the curare, we reasoned it might be the result of the effect of the prolonged artificial ventilation necessitated by the paralysis.

We turned our attention to the possibility that it was the excessive artificial ventilation that had become an integral part of the British technique that was to blame. We went on to give large doses of curare to some more cats but this time, instead of carefully controlling the ventilation at normal levels, we increased it by either an extra 50 or 100 per cent of that actually required. All these cats died. After two to six hours they showed abnormalities of their brainwaves, the heart started

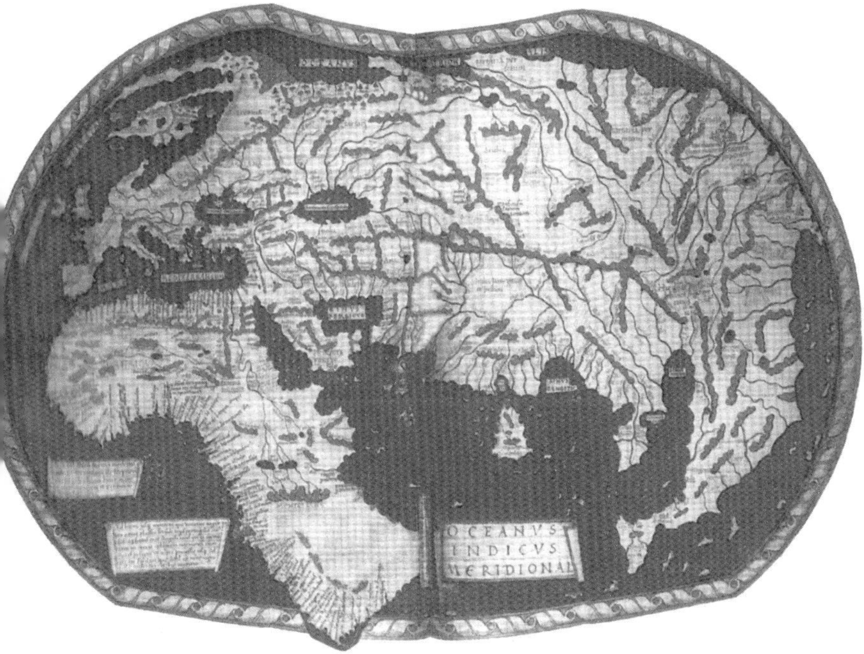

bove: A 15th-century map of the world. *(Figure 1, P5)*

elow: European adventurers arriving in South America, 'greeted'
y natives. *(Figure 2, P10)*

Charles Waterton (1782-1865), pioneering naturalist, explorer and traveller (by C. W. Peale, *National Portrait Gallery*) *(Fig 5, P19)*

Sir Benjamin Brodie (1783-1862), English physiologist and surgeon.
(*Courtesy Royal Society*) *(Fig 6, P25)*

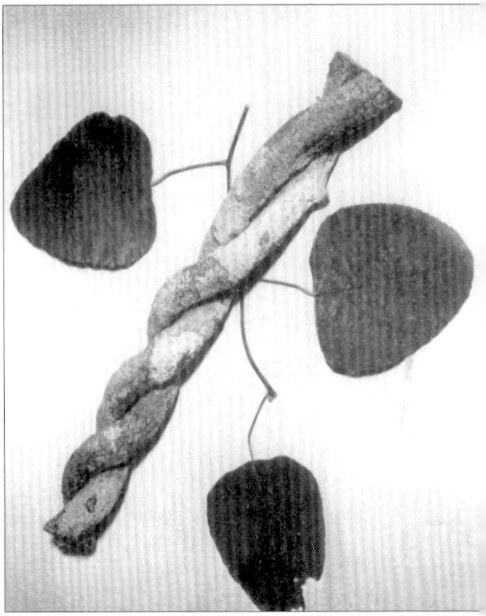

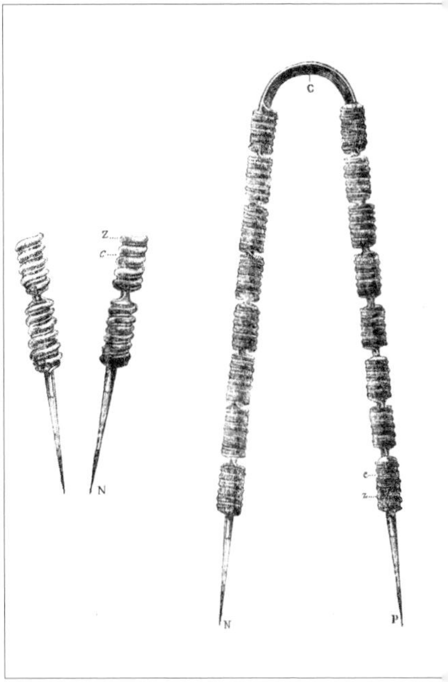

Above left: Drawing of Strychnos toxifera found in the Orinoco basin. *(Fig 7, P29)*

Above right: Chondrodendron tomentosum, the creeper found in the westerly jungles of Ecuador and Peru. *(Fig 8, P30)*

Below left: The quiver and arrows for a blowpipe. *(Fig 3, P11)*

Below right: The electrified tweezers used by Claude Bernard to stimulate nerves whilst investigating the action of curare on experimental animals. *(Fig 10, P47)*

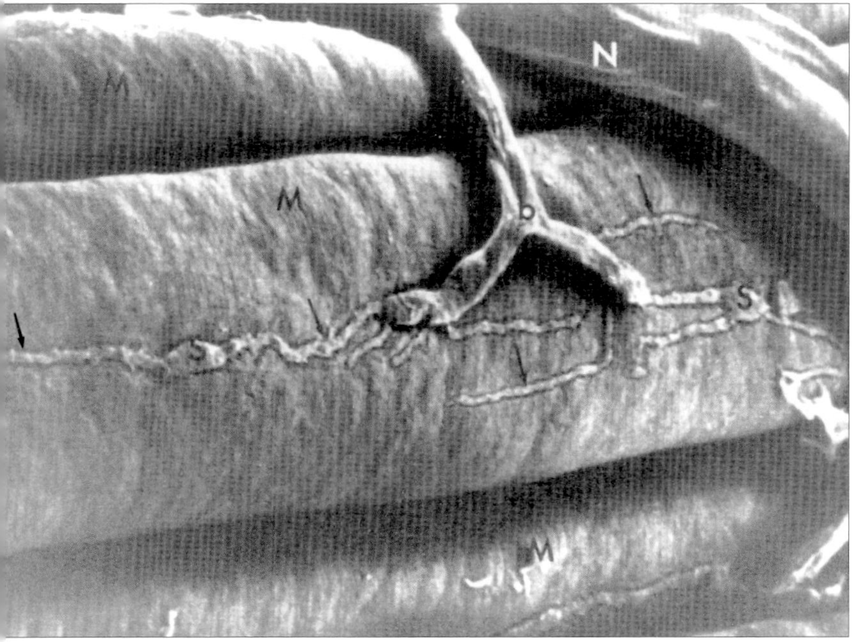

bove: Claude Bernard demonstrating to his pupils. *(Figure 11, P49)*

elow: A motor nerve ending connected to muscle fibres. Bernard emonstrated that curare acts on the nerve, not on the actual muscles. *'igure 12, P57)*

Above left: French physiologist Claude Bernard (1813-1878), circa 1866. *(Fig 9, P33)*

Above right: Dr Thomas Spencer Wells (1818-1897). (*Courtesy Royal Society of Medicine*) *(Fig 13, P72)*

Below: Natives of Ecuador preparing curare. *(Fig 4, P11)*

Sir Henry Dale (1875-1968), English pharmacologist. *(Fig 14, P95)*

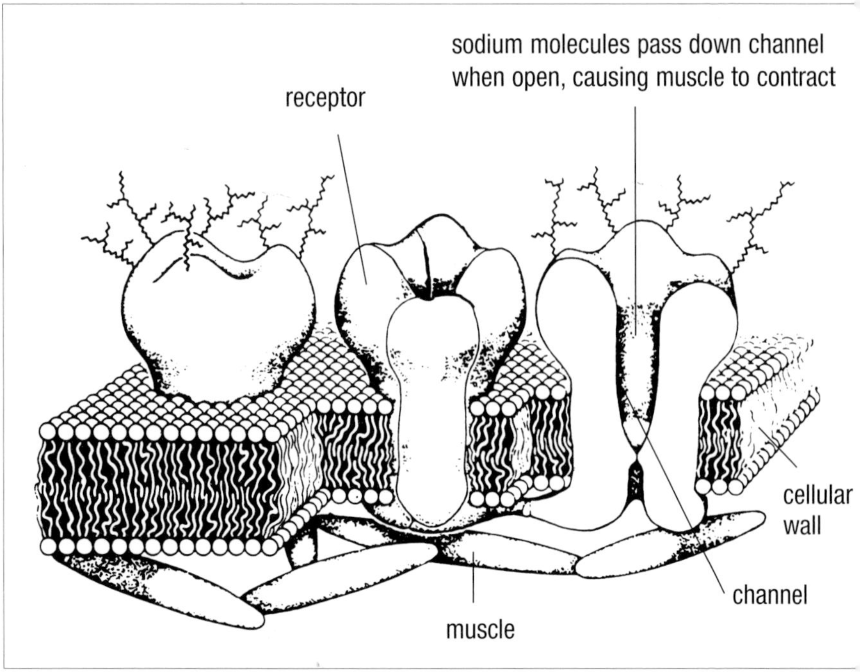

Above: Diagram of receptors. Acetylcholine binds to special sites on the receptors causing them to contract. This causes the central channel to open and sodium to pass through which causes the muscle to contract. *(Fig 15, P107)*

Below: A mongoose devouring the highly-poisonous banded krait snake. *(Fig 16, P197)*

missing beats and beating irregularly and the blood pressure fell.

What was interesting was that, as in patients with neostigmine-resistant curarisation, the fall in blood pressure did not respond to infusions of adrenaline. These experiments convinced us that it was not the curare that was at fault but excessive, poorly controlled ventilation in sick patients.

The final piece of evidence that brought it all together occurred after my return to England in 1958. These were the early days of open-heart surgery. A syndrome similar to that described by Hunter was seen after the use of the early, inefficient oxygenators that replaced the lungs while the patient's own heart and lungs were being bypassed. A biochemist, David Brooks, working with our team, showed that this was associated with an excessive amount of lactic acid in the blood due to the cellular hypoxia. It has been shown that, if there is insufficient oxygen to meet the demands of metabolism, lactic acid forms in large amounts inside the cells of the body and leaks out into the blood.

It seemed likely that the victims of neostigmine-resistant curarisation, the cats that were overventilated and those subjected to a low oxygen supply during heart surgery, all developed the same condition and that this was due to too much acid in their cells. It is evident that the various metabolic processes of the body will function properly only over a limited range of acidity in the cells and that too much acid causes them to malfunction.

We concluded that it was the excess acidity that caused the syndrome. Excessive amounts of acid or alkali would be likely to prevent the normal functioning of the cell by interfering with its metabolism. This was foreseen by Claude Bernard in the mid-nineteenth century, who clearly stated that the aim of all the controlling influences in the body is to keep the internal environment within constant, narrow parameters.

The acidity of the body fluids depends on the ratio of acid to alkali in the blood. It is probable that there was a considerable loss of alkali in Hunter's patients due to their prolonged loss of the very alkaline bowel secretions, whereas in the heart-surgery patients too much acid had been produced. The solution seemed obvious: both sets of patients needed extra alkali. We gave patients additional alkali intravenously in the form of sodium bicarbonate – baking soda. The result was dramatic. We no longer saw this problem in our cardiac patients and the syndrome of neostigmine-resistant curarisation was no longer a problem.

Acid/base balance

In our studies on neostigmine-resistant curarisation we had demonstrated that the syndrome was due to a disturbance of the all-important ratio of the amount of acid-forming carbon dioxide and the amount of base, composed mainly of bicarbonate. This balance is essential if the body fluids are to be kept in an acceptable, viable range. In neostigmine-resistant curarisation it is the loss of the base, in the form of bicarbonate, that is the problem.

This necessitates a vast increase in respiratory effort to reduce the acid-forming carbon dioxide to a point where it would compensate for the low level of base, in order to keep the ratio between the acid and base constant.

After curare and at the end of an abdominal operation, sick patients are just too weak to make the necessary ventilatory effort in order to achieve compensation. As a result, the level of the acid-forming carbon dioxide becomes too high to match the much-reduced bicarbonate concentration. This causes the acidity of the body fluids to increase to a point where cell function is impaired.

The use of curare had thrown light on the narrow parameters within which cells can function. As a result of this and other studies on the effects of artificial ventilation, we now appreciate the importance of adjusting the amount of ventilation carefully to avoid washing out too much acid-forming carbon dioxide from the blood. Maintaining the correct level of carbon dioxide is important for the efficient functioning of the body (anyone who has suffered 'mountain sickness', due to a rapid ascent to a high altitude, is feeling the effects of a reduced level of CO_2 in his or her blood). Today, with our ever-present awareness of the importance of keeping the acidity of the body fluid within narrow parameters, neostigmine-resistant curarisation is not seen.

Awareness during anaesthesia

Before the advent of muscle-relaxant drugs, patients under general anaesthesia were always asleep and unaware of what was happening to them. In the 1980s reports of patients being awake or semiconscious, but paralysed, during operations began to appear in the European journals. Often, these occurrences were the result of error, such as a faulty anaesthetic machine or failure to notice that all the anaesthetic agent had been used up.

Over the next few years these reports became increasingly common and alarming. More and more patients told of snatches of conversations they had overheard while they were being operated on. Although only a small number complained of feeling the sort of extreme pain one might have expected, there is little doubt that the experience caused them considerable distress.

To understand the background to these events, one has to review the way curare was introduced into anaesthesia. In America, Stuart Cullen's teaching that muscle relaxants are not anaesthetic agents reminded anaesthetists always to anaesthetise their patients before giving a curare-like drug. Awareness during anaesthesia was uncommon in the USA. In contrast, in England, the ability to use very light planes of anaesthesia when the patient was paralysed was seen as one of the major advantages of their use. The technique advocated by Cecil Gray and the Liverpool school deliberately used overventilation to control the breathing and make the curare 'go further'. In the 1960s and 1970s we started to appreciate the deleterious effects of prolonged overbreathing. It was found to upset the acid–base balance of the body, cause a restriction of the blood going to the brain, and in some cases it imposed a strain on the heart.

There is no doubt that the human body is better adapted to withstanding a higher level of carbon dioxide than a lower one. This overventilation washes out carbon dioxide and upsets the vital acid–base balance in the body. It is this that produces unpleasant sensations. It causes dizziness, headache and extreme fatigue. If it continues for any length of time it may cause loss of memory.

When it was realised that high levels of artificial ventilation might be harmful the technique went out of fashion. At the same time there was a movement, especially in continental Europe, to replace the volatile, smelly, anaesthetic agents with newer drugs with powerful analgesic properties, which could be given intravenously. Unfortunately with these intravenous techniques it was difficult to judge whether the patient was truly asleep. It was the combination of abandoning high levels of ventilation – which, by starving the brain of its normal blood flow, helped produce unconsciousness – and the replacement of volatile anaesthetic agents with intravenous narcotics that produced so many cases of 'awareness during anaesthesia'.

Failure of reversal

Other uncommon complications have been described with the muscle-relaxant drugs. Allergic reactions, although rare, do occur and with one particular drug catastrophic anaphylaxis and an extreme rise in body temperature associated with muscle degeneration has been described. However, although rare, the greatest fear when using a muscle relaxant drug is that the paralysis it produces will be impossible to reverse completely at the end of an operation.

According to the competition theory, on which we base our understanding of the way drugs like curare work, it should always be possible to overcome its effects by a suitable increase in the level of acetylcholine. Since acetylcholine is produced when a nerve is stimulated, preventing its destruction by means of an anticholinesterase such as neostigmine, should always be capable of reversing the neuromuscular block. The lower the blood level of drug, the easier it should be to reverse its action.

Today we refer to curare and similar drugs as *competitive blocking agents*. The demonstration that residual paralysis caused by curare could be reversed by drugs such as neostigmine removed the last obstacle to the safe use of curare in anaesthetic practice.

Although the competition theory of action forms the basis of our understanding of the way in which curare-like drugs work and how their actions can be reversed by anticholinesterase drugs such as neostigmine, it is based on work carried out in experiments where the concentration of drug remains constant during the experiment. It does not answer all the puzzles associated with these drugs *in vivo*. In clinical practice, the concentration does not remain constant as it does in the experiments in the water bath, but peaks shortly after it has been injected and declines rapidly as it is redistributed around the organs of the body.

However, there are times when it has appeared impossible to reverse a residual block even when the levels of drug in the blood are low, and insufficient to cause total paralysis. The shortcomings in our interpretation of the competition theory is borne out by two unrelated and seemingly inexplicable events that occurred following the use of curare-like drugs in patients in 1962 and 1963.

This was the situation in the case of Theresa Lopez. These events challenged our simple understanding of the theory

CHAPTER ELEVEN

Unanswered Questions

Theresa Lopez (her real name has been changed to preserve confidentiality) died on 13 May 1961. The death certificate recorded her death as being due to kidney failure, secondary to peritonitis caused by a perforated gastric ulcer. She was forty-two at the time of her death. I had been closely involved in the treatment of Theresa Lopez from the time of her admission to a London hospital, twelve days before she died. Therasa was brought into the hospital in considerable pain, as an emergency. Unfortunately, she spoke no English. She had been born in Portugal and she had been in England less than a month, working as a mother's help for a Spanish couple, before she became ill. Her duties consisted principally of looking after two children and doing light domestic work around the house. According to her employers, the sea crossing from Portugal had debilitated her and left her 'under the weather'. She could not 'get on' with English food, which she said gave her indigestion.

Some forty-eight hours before her admission to the hospital she took to her bed, looking pale and sweaty, complaining of severe abdominal pain, which had been temporally relieved by indigestion tablets. She had refused all solid food and had drunk only a little water. Twelve hours or so before her admission to hospital, she took a turn for the worse, becoming deathly pale and faint, although her pain was less intense. The doctor who was called diagnosed peritonitis.

The admitting casualty doctor at the hospital could not obtain a satisfactory history of her illness because of her lack of English, but it was obvious that Theresa was very ill. Any movement was painful and she had a rapid, thready pulse and low blood pressure. On examining her abdomen, he had no difficulty in recognising the evidence of peritonitis. Her abdominal muscles were rigid and the bowel was ominously silent. A saline intravenous drip was started and she was prepared for emergency surgery.

Theresa was operated on about three hours after her admission. At the time of her anaesthetic she had received about 2 litres of fluid, intravenously. Although her blood pressure was still low, her pulse was less rapid and her condition did not give rise to undue concern. The anaesthetic was routine and uneventful. She was given the synthetic curare-like drug gallamine to relax the boardlike rigidity of her abdominal muscles so as to allow the surgeon to explore her abdomen and determine the cause of the peritonitis.

Gallamine was chosen, in preference to curare because it was felt it would better support her blood pressure. Once the abdomen was opened it was obvious that she had perforated a gastric ulcer some days earlier, causing the peritoneum – the lining of the abdominal cavity – to become badly soiled and inflamed. The defect in the stomach was identified and dealt with by the surgeon.

The operation lasted about ninety minutes. As the abdominal wound was being sewn up, the anticholinesterase drug neostigmine was administered to reverse any residual effect of the gallamine. After a minute or so the patient started breathing, as expected, but her respirations were not as full or deep as one would have anticipated. The anaesthetic was lightened and soon the patient started to move her hands and legs, but her breathing was still laboured and not really deep enough for safety.

The period at the end of an operation is always an anxious time. The patient is at her or his most vulnerable and the anaesthetist likes to see a depth of breathing that is not only sufficient to maintain good oxygenation but also has sufficient reserve to allow the patient to cough and swallow. Good muscle tone is also necessary so that the patient can control the tongue and prevent it falling back into the mouth and blocking the passage of air in and out of the lungs.

Theresa's breathing was not good enough and it was

decided to continue ventilating her mechanically. Ten minutes later her breathing had become even more depressed and a second dose of the anticholinesterase was given. Once again, her breathing improved temporally but not sufficiently for it to be safe. Over the next ten hours the process was repeated several times, each time with the same result.

In spite of the repeated use of anticholinesterase drugs, it was impossible to reverse the residual effect of the gallamine. Each time, following the anticholinesterase, the anaesthetic was lightened and the patient started to breathe and to move her hands and legs. At times, she would grimace and try to stick out her tongue but her breathing remained too shallow for safety. Theresa's condition remained good, her blood pressure and pulse were normal and she responded to stimulation. It was decided to abandon attempts to reverse the paralysis for the time being and to ventilate her mechanically, overnight.

The next day Theresa's condition remained unchanged but it was noticed, for the first time, that she had passed no urine in spite of the infusion of a considerable volume of intravenous fluids and some 18 g of salt. In the absence of any urine it seemed probable that the difficulty in reversing the paralytic effect of the relaxant was due to her evident kidney failure. Attempts were made, over the next day or two, to reverse the residual paralysis but, each time, the result was the same.

At the same time, it became increasingly apparent that the patient's kidneys were not responding to treatment and that dialysis would be required. Some thirty minutes after starting the dialysis, Theresa was able to breathe easily, to open her eyes and to indicate that she wanted the tube that was in her mouth, through which she was being ventilated, removed. Two hours later she was sitting up in bed and talking to the nursing staff. It was obvious that reversal of her paralysis had not been possible as long as the small, residual amounts of drug remained in her blood. It was the removal of the remaining gallamine in her blood during the dialysis that made reversal of her paralysis possible.

Unfortunately, Theresa's kidneys did not respond to therapy. Fifty years ago, renal dialysis lacked present-day sophistication and kidney-transplant surgery was still at an experimental stage. In spite of continued dialysis, it proved difficult to maintain her water and salt balance. Sadly, Theresa died suddenly on 13 May 1961 during dialysis.

I was deeply disturbed by Theresa's death. Although at the time renal failure was a recognised complication of peritonitis, it is particularly unfortunate when it occurs following an underlying condition that is potentially curable. We shall never know whether the inability to reverse the effect of the gallamine contributed to her death. What puzzled me was why we could not reverse the paralysis with the antidote neostigmine.

The puzzle

It is axiomatic that the effects of a curare-like block can be terminated, either by lowering the concentration of drug or by raising the acetylcholine concentration at the neuromuscular junction. It is possible that, in the presence of very high levels of drug at the neuromuscular junction, it might not be possible to achieve sufficient acetylcholine to reverse completely the actions of a curare-like drug. However, in this case, the gallamine concentration in the blood was just sufficient to stop adequate respiration. It was obviously not sky high; indeed, it was far less than that required to cause complete paralysis. In all the laboratory experiments we had carried out, using the diaphragmatic muscle of a rat suspended in a water bath, it had always proved possible to reverse a curare-like block by increasing the acetylcholine concentration. It was a mystery why giving neostigmine, which undoubtedly increased the acetylcholine levels, did not reverse the residual, minor degree of paralysis in this patient.

Effect of kidney failure in animals

I was sufficiently puzzled by the case of Theresa Lopez that, when a few years later I had the opportunity to join Professor Ellis Cohen and his research team at Stanford in California, I jumped at the chance. Cohen had developed a radioactive method by means of which the blood concentration of the various muscle relaxants could be studied. I wanted to find out what happened to the blood levels of gallamine following administration of the drug and how it was affected by renal failure.

The research program we established involved giving a neuromuscular blocking dose of radioactive gallamine to dogs and studying the blood concentrations over the following six hours. In some of the experiments the blood vessels supplying the kidneys were tied before the gallamine was given in order to determine what contribution kidney excretion made to the lowering of the blood gallamine level. As a result of these and other experiments carried out on rats, we demonstrated that, provided the circulation to the kidneys was intact, the gallamine level in the blood fell rapidly to undetectably low levels due to redistribution of the drug to the tissues of the body, especially the liver. Within minutes of the injection, the drug could be detected in the urine. If the blood vessels supplying the kidneys were tied, cutting off any possibility of renal excretion, the blood level of gallamine still fell to about 10 per cent of its peak concentration. However, the remaining drug stayed in the circulation and its blood level did not change over next six hours, unless the ligature around the kidney was removed and renal blood flow re-established. We found that the effect of tying off the blood flow to the kidneys on the blood levels of gallamine differed significantly from the effect seen with curare. Repeating these experiments with curare showed that the blood levels of curare continued to fall, even when the blood supply to the kidneys is interrupted. It demonstrated that excretion of the drug by the kidneys is much more important in lowering the blood level of gallamine than it is with curare.

These experiments explained why Theresa remained partially paralysed after her operation. They did not solve the puzzle of why her paralysis was not reversed by neostigmine.

In a further series of experiments we used the technique Cohen had developed for autoradiography of muscle-relaxant drugs in rats. This involves giving a radioactive drug to a rat shortly before it is sacrificed. The animal is then quickly frozen, in liquid nitrogen, and set in a wax block. By exposing thin slices of the block to an X- ray plate for six days a photographic image is obtained that demonstrates the disposition of the radioactive material.

By giving radioactive gallamine or curare to rats just before they were sacrificed and using this autoradiographic technique, it was possible to find out where the drug went immediately after it was administered. It became clear from these studies that most of it was taken up rapidly, rather as a sponge takes up water, by the liver, the spleen and, provided the blood supply was intact, the kidneys.

In the case of Theresa it is probable that, after the injection of gallamine, the level of drug in the blood would have rapidly fallen as it was taken up by the cells of the liver and spleen. Ninety minutes later, at the end of the operation, the effective level of drug in the blood was so low that it was only just capable of causing a minor degree of paralysis. The unresolved puzzle remained: why, in spite of this very low level of circulating gallamine, increasing the acetylcholine by the administration of neostigmine failed to reverse the block. It is impossible to reconcile this with the convential competition theory.

The conclusion we reached as a result of these studies was that the competition theory was only part of the answer to the way curare-like drugs produce paralysis, and clearly something else is involved.

The isolated arm

The winter of 1962–3 was particularly cold. There were many days when the temperature never rose much above freezing. For days on end it was cold and snowing. Snow and ice covered the pavements around the hospital. As the hospital was situated in a very hilly part of London many accidents occurred due to patients sliding and falling on the ice as they walked up the hill.

One of the most common injuries that presented in the casualty department of the hospital was a fracture involving the bones at the wrist, a Colles fracture (named after the eighteenth- and nineteenth-century Irish surgeon Abraham Colles). These fractures are usually seen in elderly people whose bones are brittle and whose balance is impaired. But, in the winter of 1962–3, it was not unusual to see the fracture in young fit patients, too. The fracture clinic was inundated with these cases.

Attending to these patients, reducing their fractures and plastering their wrists soon became a logistical problem. One of the factors that compounded the problem was the need for them to have a general anaesthetic in order to set the fracture. A general anaesthetic requires the patient to starve for six hours before it can be given safely. This wait added to the lengthening queue of patients waiting for treatment. As a result, alternative techniques, such as local anaesthetic blocks, were frequently used.

One of the easiest and safest of these local anaesthetic blocks was a technique first described by August Bier, the father of spinal anaesthesia, in 1899. Bier's block, as it has become known, entails applying a tourniquet to the arm and injecting a large volume of very dilute local anaesthetic into one of the veins below the tourniquet. Because of the tourniquet the fluid cannot escape into the rest of the body; it is trapped. It fills the veins and distends them. The resulting back-pressure developed in the veins drives the fluid, containing the local anaesthetic, backwards into the capillary bed and into the tissues. The local anaesthetic produces anaesthesia of the limb, which lasts until the tourniquet is released. Although Bier used cocaine in his technique, newer safer local anaesthetics had led to a resurgence of this technique at this time.

Bier's blocks were used successfully for the reduction and treatment of most of the Colles fractures that came to the hospital. It provided good, pain-free conditions in most cases. Unfortunately it did not stop the muscle spasm associated with these fractures, which occasionally prevented the bones from being realigned. This was especially liable to occur in young, fit patients. In an effort to overcome this shortcoming I hit on the idea of adding a very small amount of curare to the dilute local anaesthetic we used.

I reasoned that the volume of fluid in the arm was about 10 per cent of that in the whole body and that the dose I would require would therefore be about a tenth of the usual paralysing dose of curare. I settled on adding a dose of between 2 and 3 mg, depending on the patient's size. First, I tested the effect of an injection of this dose of curare into my own veins to see what would happen if the curare escaped into the rest of the body. I found that, apart from a little short-lived blurring of vision, it had no adverse effect.

I soon had a suitable subject on whom I could test out my idea. He was a fit, twenty-eight-year-old patient with a Colles fracture with marked bony displacement. The trial worked well: his hand became anaesthetised and paralysed, and his fracture was reduced painlessly, without undue effort. I released the tourniquet on his arm, as the plaster slab was setting, after first warning him that he might have a little blurring of vision. The patient appeared perfectly happy and felt no adverse effect.

After the plaster was set and the bandage applied, the surgeon asked the patient to wriggle his fingers to make sure the bandage was not too tight. To my consternation the patient found that he could not move his fingers at all. I must have conveyed something of my anxiety to the patient who was highly amused by my discomfiture. Although he said his fingers still felt numb his reaction to pinprick made it clear that sensation had fully returned following release of the tourniquet, but his ability to move his fingers remained very limited and he interpreted this as numbness.

It was about twenty minutes after releasing the

tourniquet before there was convincing evidence that muscle power was returning. It took over forty minutes before I was happy that his muscle power had fully recovered and that he was fit to leave.

Although pleased that my test had worked, I was amazed and puzzled that the curare effect did not wear off as quickly as that of the local anaesthetic drug once the tourniquet had been let down and fresh blood, free of any drug, flooded into the arm.

I had not been unduly worried by this event, as I was certain that full recovery would take place. What puzzled me was why, when the arm was being perfused with fresh blood that contained no curare, its effect lingered for about forty minutes. It is implicit in the competition theory, on which we based our understanding of the way curare-like drugs work, that the amount of block depended on the concentration of drug relative to that of acetylcholine. Clearly the concentration of curare left in the blood in the arm after the release of the tourniquet was tiny. Its level in the circulation was insufficient to cause any weakness, even in the most sensitive muscles, yet the block remained.

It seemed highly unlikely that the brief period for which the tourniquet had been applied would have been sufficient to affect the production of acetylcholine since, in the absence of curare, a Bier's block did not cause any muscle weakness. I determined that, before I repeated the experience, I would learn more about what happened when muscle-relaxant drugs were used in the 'isolated arm'.

It was these two observations, one in the case of Theresa Lopez and the other in the isolated arm, that aroused my curiosity about the way muscle-relaxant drugs work in patients. The competition theory fitted all the results of our experiments carried out on the rat diaphragm in water baths in the laboratory, but it was difficult to reconcile these findings with what had occurred in the patient with the Colles fracture and in the case of Theresa Lopez.

It was these two events that caused me to seek an answer. It was a puzzle that had to be solved. It was the start of several years of experimentation.

CHAPTER TWELVE

The Way Forward

'As to complete certainty, no man has known it.'

DESOPHONES

'There should be no such thing as consensus in science. The greatest science is when one breaks with consensus.'

MICHAEL CRICHTON, AUTHOR

To collect the evidence to support any scientific theory requires a great deal of detective work. It is only when all the evidence points to the same conclusion that the theory becomes a credible hypothesis. As in a murder enquiry, it is necessary to collect the clues, piece by piece, to test their accuracy and piece them together so that an overall pattern emerges. No single clue can be construed as proof in itself. It is the accumulation of evidence, all of which points in a particular direction, that provides the confidence necessary to support any hypothesis.

If any evidence arises that suggests you are on the wrong track, for example an alibi that makes it impossible for the suspect to have been in the country at the time of the crime, then, provided this is proven to be correct, it means that you have got the theory wrong. Disproof is a more powerful investigative weapon than proof.

As in a detective investigation, circumstantial evidence can help to build up a case in favour of a particular theory but it does not constitute proof. Proof comes only from evidence; even then, one must exercise caution, for, as the Greek philosopher Desophones pointed out, science deals with the 'most probable' rather than the 'absolutely certain'.

I was puzzled by the failure of the antidote to curare, neostigmine, to reverse the paralysis in the case of Theresa Lopez. It is true it was an unusual, possibly unique occurrence, but it could not be explained by the competition theory. If such circumstances could be reproduced, it would constitute 'disproof'.

This feeling was reinforced by what happened to the patient whose fractured wrist I had anaesthetised using a Bier's block containing local anaesthetic together with a small dose of curare. The failure of the paralysis to wear off immediately when the tourniquet was released could not be explained by the competition theory. That his muscles did not recover immediately following the release of the tourniquet and recovery from the effect of the local anaesthetic suggested other factors were at work.

If we were to understand how curare-like drugs work in patients in all circumstances, then it is essential to understand how events like this could happen at all, even if they were unusual. This was the start of a series of investigations that covered almost twenty years. As in any forensic examination, evidence was collected, piece by piece, until it pointed us towards a possible explanation.

The competition theory is the universally accepted hypothesis explaining the way in which curare-like drugs cause neuromuscular block. It is supported by a wealth of laboratory experimentation. It proposes that the molecules of a drug like curare are in constant competition with molecules of acetylcholine for a momentary occupancy of a site on the receptor where, in a matter of milliseconds, it reacts to cause a response. For each muscle fibre this is an all-or-none affair, in which the winner takes all. Either acetylcholine wins and there is no block or, if curare wins, a block occurs. It is like a prize fight where only a knockout determines who has won.

All the investigators working on this problem in laboratory conditions produced results that more or less fitted this concept. The majority of these studies, like our own, were made on a standard research preparation. A rat's diaphragm is removed from the animal and suspended, together with its nerve, in a water bath containing a given concentration of curare. Increasing concentrations of acetylcholine-like drugs are then added to the water bath and the effect on the contraction of muscle is measured following stimulation of its nerve supply. This type of *in vitro* experiment confirms a quantitative relationship between the concentration of curare in the water bath and the dose of acetylcholine needed to reverse its paralysis. Small differences between the competitiveness of a particular drug with acetylcholine for the receptor can be demonstrated, but, as with pitting boxers of differing skills against the same, well-trained adversary, they do not undermine the basic competition hypothesis.

The problem

It was against this background that we sought to explain why, in the case of Theresa Lopez, the antidote to the drug did not reverse the minor degree of neuromuscular block that was present. In the Bier's-block case, in the isolated arm of the patient with a Colles fracture, the block continued long after the curare had disappeared from the blood. Neither of these two separate events can be readily explained by the competition theory.

It was impossible to reproduce the clinical situation present in the case of Theresa Lopez but we could easily repeat, in volunteers, the conditions of the isolated arm. This was the obvious way to start our investigation to try to find an answer to the puzzle. The first thing to do was to try the isolated-arm experiment on oneself to see if the effect was reproducible.

I needed to persuade a colleague to help. While it was possible to inject oneself safely with a tiny dose of dilute curare, it would have been irresponsible to replicate the Bier's block without a competent person at hand, should things go wrong. One of my colleagues, John Hoyle, volunteered to help me with the experiment.

The very next day we repeated the conditions of the isolated-arm block using my right arm as the test bed. An inflatable tourniquet was placed on my right arm in the same way that a doctor puts a cuff on your arm to take your blood pressure. An intravenous cannula was inserted into a vein on the back of my hand. The tourniquet was then inflated so as to stop any blood entering or leaving my arm. Two syringe-fulls (40 ml) of saline containing 3 mg of curare was then injected, as quickly as possible, through the cannula into the vein on the back of my hand. For the purpose of these experiments it was not felt to be necessary to use any local anaesthetic in the saline since the presence of a second drug might complicate the interpretation of the results. We deliberately used a high dose, 3 mg, of curare to try to maximise any effect.

The first surprise was that injecting the saline rapidly was quite uncomfortable. The distended veins became painful and tense. I then experienced the sensation that my patient with the Colles fracture had described: that the hand felt numb even though there was no local anaesthetic in the injected saline. Slowly, the muscle paralysis developed and, after three minutes, movement in the hand became impossible.

At this point the tourniquet was released. As fresh blood flowed into the arm I developed transient double vision and difficulty in focusing clearly. Clearly, some of the curare had been washed out of the arm and was circulating in the bloodstream and had been carried to the muscles of the eye. The double vision was not surprising, as it is well known that the eye muscles, especially those that coordinate the vision of the two eyes, are very sensitive to curare.

The effect wore off rapidly. Three minutes after the release of the tourniquet I was able to read comfortably, move about normally and use my left hand. The only part of my body affected by the curare was my right hand, and that was completely paralysed. By twenty-five minutes the muscle power was starting to return to the right hand. At forty minutes I thought the function of my hand had become normal, but trying to do up the buttons on my shirt proved impossible, because my hand movements were still too clumsy to control delicate muscle activity. It was a further twenty minutes before I could use my right hand to hold a pen and to write.

Clearly, the experience of the patient with the Colles fracture was reproducible. Up to sixty minutes after the tourniquet was released and the curare level in the circulating blood was reduced to a level that had no effect in the rest of my body, a considerable paralysis was still present in my right hand. Lowering the drug concentration in the blood far below what would cause a neuromuscular block had not reversed the paralysis as it should if the competition theory was right. It was as if the drug had become stuck on the muscle receptor during the three minutes that the tourniquet had been inflated and had remained there, in a sufficiently high concentration to cause a neuromuscular block long after it had been released.

Without measurement, science lacks the objective evidence it needs for proof. It was necessary to find some means of measuring the effect of the curare in the arm, isolated by the tourniquet, and comparing it with the strength of the muscle contraction of the other, non-isolated, arm. Our first attempt to measure muscle strength in the hand was by means of a rubber bulb placed in the hand and recording the pressure produced when it was squeezed, by means of a manometer filled with mercury. When we used this crude apparatus to test the effect of curare in the isolated arm, we found that we could demonstrate muscle weakness for about thirty minutes following release of the tourniquet. The duration of this effect varied from time to time, but was seldom less than twenty minutes. Our results were too diverse to be convincing. It was clear that we needed to refine our technique and to produce a reliable record of the effect of the drugs tested on muscle power if we were to convince others.

A newly appointed research fellow, Malcolm Tyrell, joined me in these studies. He approached our Department of Clinical Measurement for help in designing an apparatus to study muscle power in the isolated-arm experiment. It was decided that the simplest method would be to monitor the force of contraction of the muscles at the base of the thumb in response to stimulation of the nerve supplying them. The adductor muscles, as they are called, cause the thumb to move across the palm of the hand in response to nerve stimulation.

They come into play when one is gripping an object, such as a pen.

These muscles are known to have been very important in the evolution of intelligence. They gave hominoids the ability to pick up and hold an object in their hands so that it can be examined and its function explored. The direction of the movement that occurs when their nerve is stimulated is in a different direction from that produced by the other muscles, which act to close the fingers onto the palm. This allows the response of the muscles of the thumb to nervous stimulation to be measured separately from those of the rest of the hand. Their use in monitoring neuromuscular block had been suggested some years earlier by the New York anaesthetist Ronald Katz.

The other reason for selecting these muscles was that the nerve supplying them enters the hand on the opposite side to the thumb. It is most easily stimulated on the little-finger side of the wrist. Because the point of stimulation is so far away from the affected muscle there is very little chance of the electrical impulse being conducted directly to the muscle, a frequent cause of error in this type of investigation. The Department of Clinical Measurement rose to the challenge and produced a sensitive but expensive monitor using a strain gauge. A strain gauge produces a fall in electrical resistance, which can be measured easily, in response to pressure. Any change in resistance can then be recorded. The apparatus was mounted in a bicycle handgrip to allow it to be fixed firmly in the hand. All the early experiments were carried out with this apparatus, but we soon discarded the expensive, customised strain gauge for a cheap alternative produced commercially for use in bathroom scales. In our first series of tests we regularised the technique in an effort to produce more consistent results. We decided always to use the same volume of saline, to keep the tourniquet inflated for three minutes and to use a dose of relaxant that was one-tenth of that causing paralysis in average-size man.

The isolated-arm experiments

The first drugs we tested were curare and gallamine. I deliberately chose to use gallamine because I was anxious to shed light on the effect I had seen in Theresa Lopez and hoped that this experiment would go some way to providing an answer.

The recovery of muscle power in the isolated arm after gallamine proved to be consistently quicker than when curare was used. This difference in the recovery times was very similar to that found when a fully paralytic dose was given to anaesthetised patients. In both the isolated-arm experiments and in the anaesthetised patient, the paralytic effect of curare lasted about 15–20 per cent longer than gallamine. The consistent difference in the duration of action of these two drugs in the isolated arm could not be due to a different concentration of each drug in the blood, since, once the tourniquet was released, the blood perfusing the paralysed arm after both the gallamine and the curare would have been negligible; it was much too small to produce any effect on other muscles.

The results of these early experiments led us to the inevitable conclusion that, following administration of the drug in the isolated arm, a large proportion of the molecules became fixed at the receptor, where they remained even after the tourniquet was released and fresh blood containing no drug circulated through the arm. The duration of this 'fixation' period appeared to be determined by the chemical structure of the drug. The paralytic effect of the drug in the isolated arm clearly was not determined by its relative level of drug to blood, as the competition theory would lead us to expect.

We would have liked to have been able to study the effect of acetylcholine on the recovery rate of different drugs after their injection into the isolated arm; unfortunately, this would have been unsafe. Acetylcholine-like drugs affect many bodily processes besides the neuromuscular junction, including the heart, the salivary glands and the bowels. Although some of these effects can be minimised by pretreatment with atropine, the effect on the gut cannot be completely prevented. No volunteer was willing to suffer the severe bowel cramps that were likely to occur; even I balked at the idea.

As a result we had to use a less satisfactory means of increasing acetylcholine levels locally during the recovery period. It is known that the faster the rate of nerve stimulation the greater the output of acetylcholine. This is due to the activation of a feedback mechanism. In order to achieve the same effect as increasing the acetylcholine level we chose to use bursts of high-speed stimulation of the nerve to the thumb. We turned up the rate of stimulation from two to thirty times a second. As this can be painful, we applied these bursts of stimulations only at minute intervals.

We showed that increasing acetylcholine in this way reduced the time it took for recovery of a curare block by about 15–20 per cent. It is of interest that, way back in 1938, Bennett, who used Intocostrin to prevent the major, bone-shattering convulsions of 'shock therapy', had observed that the convulsions that occurred hastened the recovery of muscle power. Undoubtedly, the Metrazol he used to produce convulsions would have caused a sudden, huge release of acetylcholine.

In our experiments, acetylcholine appeared to help prise the curare off the receptor, releasing it and so speeding up recovery. If this effect had been the result of the increased release of acetylcholine temporarily competing with the drug, its effect should have stopped once the burst of stimulation had ended. However, as in Bennett's patients, the recovery continued long after the extra acetylcholine was gone. It is difficult to reconcile this effect with the competition theory.

Our findings were presented at a meeting of the Royal Society of Medicine in 1970. They were greeted with disbelief. It is always difficult to persuade a consensus of those steeped in a particular way of thinking that they may be wrong or that, although the theory might be correct, their interpretation of it may be incorrect. This was evident from the reluctance to abandon Sherrington's theory of 'electrical transmission of nervous activity' in the early part of the century. The establishment was too committed to the competition theory to consider any alternative, especially one that was based on such a highly unusual method of administering the drugs.

Our findings were dismissed as being due to an overdose, an effect of the anoxia produced by the tourniquet or an unusually enhanced effect of the small, subparalytic amounts of circulating drug. Only those who had actually witnessed the experiments paused to reflect on the important implications of our findings. The sceptical opinions were strengthened by new research from America, which, at first sight, offered strong support for the competition theory.

Measuring blood concentrations

Although there had been many attempts to devise a method of analysing the blood levels of curare, they had either been too difficult to replicate consistently or, like Ellis Cohen's technique, which we used in our autoradiographic studies, required the use of radioactive drug, a technique that could not be justified in man.

In 1974, R. S. Matteo, S. Spector and P. E. Horowitz described a radio-immune assay that was suitable for accurately determining the curare levels in the blood of patients. Using this method they were able to demonstrate that, following an injection of a paralysing dose of curare, there was a rapid rise in its concentration in the blood until it peaked; this was initially followed by a very rapid fall. When the drug concentration had fallen to about 10 per cent of its peak concentration, the rate of decline flattened off, becoming slower and slower, until it reached levels too low to be measured.

This biphasic effect is due to the initial rapid redistribution of the drug to the cells of the liver, kidneys and muscles, followed by a longer, slower decline in blood level as the drug is slowly excreted from the body by the liver and kidneys. At the same time as they were sampling the patient's blood, they studied the effect of the curare on the patient's muscle power. They found that the paralysis occurred rapidly following injection of the drug and the rise in its plasma concentration. As the concentration of drug in the blood fell from its peak level, the muscle power started to recover at a rate that was roughly in concert with the blood level of drug. They concluded that the correlation they had shown between the plasma level of curare and the recovery from neuromuscular block proved that it was the concentration of drug in the blood that controlled the degree of the block. This finding is what one would expect on the basis of the competition theory.

Their conclusion appeared to be in direct contrast to our findings. We had demonstrated, in the isolated-arm experiments, that a curare-like block continued for up to forty minutes after the blood level of drug had fallen to levels that were too low to measure, whereas their work suggested that the rate of decline in the blood concentration and the recovery from paralysis were in step with each other.

Obviously, we could not both be right. I suggested that Matteo and his coworkers had read too much into their findings. They had demonstrated a casual relationship between the blood level of curare and the amount of paralysis, but this was very different from proof of a causal relationship. While recovery would clearly have been impossible without the plasma level falling below that necessary to produce paralysis, their results did not show that it was the rate of the fall in blood concentration of drug that had determined the rate of recovery. The interpretation of a relationship between two events as cause and effect is a common cause of confusion. It is perfectly possible that the initial steep fall in plasma concentration of drug they had demonstrated had reduced the curare concentration below that necessary to produce paralysis, just as the release of the tourniquet did in the isolated arm, and had nothing to do with the slow recovery from the curare block. In that case the slow recovery they had demonstrated might not have been the result of the fall in blood level of drug but to the slow release of curare that had been stuck at the receptor. They had failed to demonstrate a truly causal relationship. The relationship they found between plasma level and paralysis did not constitute scientifically valid proof that one was dependent on the other.

This may seem an academic, recherché argument but it is one that was of immense importance. It had to be resolved if we were to accept the competition theory.

One example of the importance of resolving this

dispute was the way it influenced the search for a new, short-acting drug suitable for administration for short operations. The conviction that it was the plasma level of drug that determined the degree of block gave the impetus to one of the most expensive research programmes in this field. The firmly held view of the team working at Harvard was that, if they produced a drug that was rapidly destroyed in the blood, it would be eliminated from the plasma in seconds and it would therefore, of necessity, be very short-acting. This was a reasonable supposition if one accepted the competition theory.

On the basis of our results in the isolated arm, we predicted that, even if they produced such a drug, the short action they wanted would not necessarily follow if the chemistry of the drug was such that it became firmly stuck at the receptor. It took the Harvard team about ten years and a huge research expenditure to produce such a drug. It was mivacurium. It was so rapidly destroyed in the blood that its concentration fell almost as quickly in the patient as curare does after release of the tourniquet in our isolated-arm experiment. When it was finally tested on patients, it proved, as we had predicted, not to be short-acting. Instead its duration of action was similar to that of gallamine, a muscle relaxant used in anaesthesiology. When it was tried out in the isolated arm, we found its duration of effect was also similar to that of gallamine. The drug continued to cause neuromuscular block long after it had completely disappeared from the bloodstream in the isolated arm. It proved to have a similar duration of action in clinical practice.

Even if one accepts the idea that a drug becomes stuck at the receptor with a firmness of binding that is dependent on its chemical structure, it has to be recognised that there can be no recovery from the effect of the drug if the blood level remains high. If a relatively high level of drug is maintained in the circulation, there is no prospect of its being removed from the receptor. There must be a concentration gradient between the receptor and the blood if drug is to be able to leave the receptor. It would be anticipated that the greater this gradient, the easier and faster would be the recovery of neuromuscular transmission. The concentration of drug in the blood must play some part in determining the rate of recovery from paralysis, even if it is not the controlling factor. In order to clear up this confusion we needed to find a way of determining how the blood level affected the duration of action of curare-like drugs.

The role of the plasma concentration

Following an International Muscle Relaxant Conference, I discussed this problem with two fellow research workers. One of them was Sandor Agoston, one of the most engaging and intelligent figures in the muscle-relaxant field. He had recently developed a chemical method for determining the concentration of the new drug pancuronium in the blood and was anxious to try it out in a research project. The other was Professor Ron Miller, from San Francisco, who was involved in studies on the uptake of muscle relaxants in the body and their excretion by the kidneys.

The problem, as we saw it, was to determine whether it was the blood level or the chemical properties of the drug that determined how long a block was produced. In a series of experiments carried out in three centres, one in the UK, one in Holland and the other in San Francisco, we established that the quickest recovery from a neuromuscular block occurred when the drug level in the blood was lowest. Conversely, in the presence of significant levels of drug in the blood, even if this were insufficient itself to cause paralysis, spontaneous recovery from neuromuscular block was prolonged. The higher the blood level, the lower the concentration gradient between receptor and blood, the longer it took for recovery to occur.

Here at least was a partial explanation of the problem of Theresa Lopez. It seemed likely that, because of the failure of her kidneys to excrete the drug she had been given, she was left with a small, but significant, blood level of drug at the end of the operation. She would have been in a similar position to the patients in our study, who, at the end of a prolonged infusion, had higher than the usual drug concentration in their blood. We found that these patients took a very long time to recover normal neuromuscular function.

In the patients who had a prolonged infusion, however, excretion of the drug continued during the recovery period, lowering its concentration in the blood. This allowed reversal of the block eventually to take place.

Unfortunately, in the case of Theresa Lopez, this could not happen because of the renal failure. As a result, when I tried to reverse the paralysis with anticholinesterase drugs, any gallamine displaced from the receptor by the action of acetylcholine would not be able to leave the receptor area, as the blood had too much drug in it for it to be easily washed away. Once the effect of the anticholinesterase waned and the local concentration of acetylcholine fell, molecules of the drug would once again become bound at the receptor.

One can look on the receptor as being like sticky flypaper and the molecules of drug as flies. Anticholinesterase drugs would cause the flypaper temporarily to lose its stickiness. When this happens flies would be free to escape and leave the room. Once the flypaper became sticky again it would be likely to remain virtually fly-free if the flies had flown away. However, if the flies could not escape and remained in the room, it is inevitable that many of them would become stuck once again. Although this explanation seems reasonable and would explain many of our observations, it depends on the molecules of drug being 'sticky'. In pharmacological terms, the molecules have to posses an 'affinity' with the receptor. This does not fit easily with the idea of competition between freely available, mobile molecules.

A brainwave

It was at this time that my new research fellow, Nick Fauvel, had a brainwave. Like all good ideas, it was simple and obvious. We were suggesting that curare-like drugs had a degree of stickiness, which varied from drug to drug according to its chemistry. It was our proposal of a particular 'stickiness' of drugs that put us in conflict with the competition theory. If we could show that there was a consistent difference between two different drugs in the same person at the same time, then it would provide strong support for our idea.

Fauvel suggested we could do this simply by putting the tourniquet on the forearm rather than the upper arm of our volunteers. If we did this we would need only half as much drug to produce the same effect and we could then use both arms, simultaneously, each being tested with a different drug, in the same volunteer. The big advantage of this simple extension of our technique was that it allowed us to use a control drug in one arm and a test drug in the other, while both arms would be perfused by the same blood once the tourniquet was released. This obviated any possibility that the result would be influenced by the tiny concentration of either drug in the blood.

The idea worked brilliantly. We could demonstrate time and time again that certain drugs recovered more quickly than others in the same individual. Some of the most interesting studies were performed using two chemically similar, steroid-based relaxant drugs, pancuronium and its newer, shorter-acting sister drug, vecuronium. Both these drug molecules were of similar size and structure. Pancuronium neuromuscular block invariably took about 25 per cent longer to recover than vecuronium.

With all the curare-like drugs tested, the recovery rate in the isolated arm paralleled that in found in clinical practice when a full dose of drug is administered to an anaesthetised patient. This supports the view that the recovery rate from curare-like muscle relaxants is determined primarily by its chemical structure and, in particular, by its electrostatic charge, in both the isolated-arm experiments and during clinical anaesthesia.

We looked for other ways of supporting our binding hypothesis so that we could build up a dossier of evidence. Together with Dr Valerie Goat we hit on the idea of examining the effect of increasing the blood flow on the recovery time from a curare-like block. We reasoned that, if the drug were bound at the receptor, increasing the blood flow would not easily remove it, whereas, if it were free in the environment and merely competing freely with acetylcholine, it would be easily washed out.

We set about testing the effect of varying the blood flow through paralysed limbs of anaesthetised dogs. These experiments showed that even an eightfold increase in blood flow did not increase the rate of recovery from the paralysis. This was strong evidence in favour of the binding hypothesis.

The hypothesis

As a result of these and many other experiments, we proposed a hypothesis suggesting that curare-like drugs become bound at the receptor with an affinity, or 'stickiness', determined by their chemical structure. Since the binding of curare-like drugs appears to occur earlier than the onset of neuromuscular block, we suggested that the two events occurred separately. To explain this we proposed that this might be due to the binding and blocking actions of curare-like drugs occurring at two different but nearby parts of the receptor complex.

This theory allowed us to reconcile the competition theory with our observations in the isolated arm. After all, the actual part of the receptor that reacts with the single choline on the molecule of acetylcholine, the recognition site, is less than 2 per cent of the whole structure. To indicate our uncertainty as to the actual site and nature of this binding, we referred to it as the *biophase*, to indicate a vague biologically active area at the receptor.

Our present understanding of neuromuscular pharmacology owes much to Professor Bill Bowman of Strathclyde University in Scotland. He helped unravel the feedback mechanism that mobilises the reserves of acetylcholine required for sustained muscular exercise. Bowman was also puzzled by our observations. He felt they had to be accommodated within the accepted mantra of the competition theory. He accepted that there was a large body of evidence to support both concepts and that neither could easily be dismissed. He came up with an intriguing suggestion that also answered another completely separate conundrum. He tackled the question by asking himself why it was necessary for virtually all the powerful curare-like drugs to have two separate highly charged choline-like molecules, separated by a preset distance of 1 nm (nanometre, or one thousand millionth of a metre), when acetylcholine, the actual transmitter, has only one.

He suggested that perhaps each choline 'head' acted at a different, but closely associated, site at the receptor. One head would then act at the acetylcholine recognition area, where it would compete with acetylcholine to produce neuromuscular block, while the other choline group would react at an adjacent site, where it would anchor the drug by a physicochemical process whose strength was determined by its stickiness or affinity. The total stickiness of the drug would be the sum of the affinities at both these sites. An excess of acetylcholine would diminish the degree of overall stickiness, as it would weaken the attraction between the drug and the its recognition site, reducing the totality of the drug's affinity. This sort of drug-receptor reaction is believed to occur at other receptors and accounts for the long duration of action of certain drugs whose half-life in the blood is short. It would explain why, when there are two drugs so structurally similar as vecuronium and pancuronium, the drug with the lesser electrical charge on one of its choline-like molecules, which reduces its overall affinity, is shorter-acting.

Bowman's suggestion, although unsupported by firm evidence, received support from an unlikely source. It came from a study into the poisonous effects of a venomous snake, the banded krait.

CHAPTER THIRTEEN

The Mongoose and the Snake

It has been known for hundreds of years that keeping a mongoose in the house protected the inhabitants from snakes. In Africa and Asia these small furry rodents are often kept as pets. The mongoose is a natural carnivorous predator feeding on small rodents and snakes. It often appears to be playing with its prey, especially if it is a snake, like a cat with a mouse, nipping it and retreating, before closing in for the kill. (Fig 16)

During these forays, it is not unusual for the snake to land a bite on its tormentor, but this seems to have no effect on the mongoose. The mongoose seems to be totally immune to the reptile's poison. The venom of the banded krait is deadly; it kills small mammals and is a frequent cause of fatal poisoning in children and small or infirm adults. For years it has been generally assumed that the mongoose's apparent immunity to the snake's bite was due to the possession of a natural antibody to its venom, but none has ever been identified.

In the last fifty years or so, snake venoms have been studied in many centres around the world in the hope that, once the way they worked was known, it would prove possible to produce an antidote to the poison. Many of the viper venoms, together with some of poisons extracted from poisonous toads, snails and fish found in the Far East, are made of the same basic molecules as the proteins of the body. They contain short chains of amino acids, called *polypeptides*, which poison the victim by causing a paralysis. They act, like curare, by blocking the action of acetylcholine at the junction between the nerve and muscle.

The poison of the banded krait contains several of these polypeptides, two of which, alpha- and beta-bungarotoxin, act at the neuromuscular junction. Alpha-bungarotoxin sticks, on the receptor at the motor end plate, with an avidity that makes it difficult to dislodge. As a result it causes a prolonged paralysis. For this reason, radioactive alpha-bungarotoxin has been widely used as a marker to allow identification and separation of this type of acetylcholine receptor.

In spite of many attempts to find a natural antibody to the snake's poison in the mongoose, none has been identified. The puzzle remains. Why is the mongoose not killed by the venom from the banded krait?

Between 1992 to 1994, a team at the Weizmann Institute of Science in Israel investigated the cause of the mongoose's apparently immunity. Their starting point was the reptile itself. After all, snakes do not succumb to the inevitable occasional bites of other snakes, so it seemed reasonable to assume they possessed some particular way of protecting themselves from its paralytic effect.

The team started by examining the makeup of the acetylcholine receptor in the snake. Under the microscope it appeared to look like any other muscle end-plate receptor. It was composed of the same five molecular subunits arranged in the same order as in the other receptors that have been studied. It was only when they analysed the detailed chemical composition of these subunits that they found a small difference in the order of the chain of amino acids that make up the special alpha subunits.

It is these alpha subunits that contain the actual acetylcholine docking site. Even then, the alterations that they observed affected only a tiny area of these subunits. The majority of the molecule had the same composition as that in other animals.

The five subunits that make up the receptor are each elongated molecules made up of more than a hundred amino-acid moieties connected together to form a long convoluted chain. They are arranged side by side to form a crown around a central pore. They penetrate the cell membrane extending from the inside of the muscle cell to protrude, like lips, on the outside of the muscle. It is this protrusion that gives the surface of the receptor a doughnut like appearance when viewed under a powerful microscope.

It is only a tiny part of this structure, less than 2 per cent, that reacts with acetylcholine to open the central channel. It lies between the beadlike amino acids at number 128 and 142 in the sequence of the molecules in the chain. At this point there is a loop in the chain, like a cul-de-sac protruding from one side of the amino-acid chain, with the two amino acids, between 128 and 142, coming together at the neck of the loop. These two sulphur-containing amino acids are highly reactive and it is here, at the so called 'di-sulphide loop', that acetylcholine reacts with the receptor.

Sara Fuchs and her colleagues at the Weizmann Institute found this di-sulphide loop was present in the exactly the same form and position in the snake as in all other animals. About 10 amino acids away, at 1–1.2 nm further along the chain, they found a difference in the arrangement of five molecules. The more reactive sulphur-containing molecules were missing and replaced by more benign, neutral ones. These were part of four or five amino acids they found in this region that differed, either in position or chemical makeup, from those found in the receptors of other animals, such as the rat, cat and rabbit.

This abnormal area was close enough to the acetylcholine recognition zone to suggest that it played some part in the apparent inability of the snake to bind bungarotoxin. It is tempting to postulate that it is this abnormal sequence of amino acids found in the snake, where the reactive sulphur molecules were replaced by the less active ones, that reduces the snake's ability to bind the venom, alpha-bungarotoxin, and to prevent it from having more than a temporary effect on the krait.

A similar difference in the sequence of amino acids was found when they went on to investigate the mongoose receptor. In both these animals alpha-bungarotoxin does not become bound at the receptor and its effect is trivial. Although the mongoose does react when bitten by a snake, the effect is very short-lived and little more than a temporary annoyance.

In the mouse, the chick and man, the amino-acid sequence in this region is normal. The reactive, sulphur-containing amino acids are found close to the area where acetylcholine acts. In these species, autoradiographic studies using the radioactive alpha-bungarotoxin, confirm that it binds so strongly to the receptor in this region and that it is extremely difficult to remove even after repeated washing. This causes a prolonged action and lethal paralytic effect when these animals are bitten by a snake.

They concluded that it was this small difference in the sequence of the amino acids, at a site close to the acetylcholine recognition zone, that prevented the bungarotoxin in the snake's venom from binding with the receptor, and this protected both the mongoose and the snake from the poison.

These results fit in with Bowman's suggestion that, close by the region where acetylcholine reacts, there is second site where positively charged molecules, such as choline, can become strongly bound. It is proposed that it is this site that holds one of the charged choline groups on the curare molecule in place while its second choline, 1 nm away, competes with acetylcholine for the part of the receptor molecule that controls the central pore at the nearby site. It is this site that determines whether or not the 'gate' is opened and action takes place.

It is as if there were two similar magnetic poles, set close together in the cell membrane, but the power of one was considerably weaker than the other. As a result, one pole will bind an opposite magnetic charge more strongly than the other, prolonging its effect. The necessity of these drugs to act at both of these sites, at much the same time, would explain why curare-like drugs are most active when they have two choline groups, separated by a distance of 1 nm. The difference in the affinity of these two sites would explain why, in the isolated arm, the drug appeared to became fixed at the receptor about two minutes before the onset of any paralysis.

Other animals

The Weizmann team went on to show that there was no immunological difference between the alpha units in the mongoose, and the mouse and the chick, indicating that they come from the same gene pool. This strongly suggests a common evolutionary pathway for their receptors in spite of the difference in the sequence of amino acids. It points to the probability that all the acetylcholine receptors have developed from the same basic template. It would be surprising if minor deviations from this common pattern, such as they discovered, had not occurred during the millions of years it took for the different species of animals to evolve.

By studying the effect of alpha-bungarotoxin in different species it can be seen that the susceptibility to viper poison varies from animal to animal, depending on the structure of its receptor. While the mouse and the rabbit are both very sensitive to the poison, and the snake and the mongoose almost totally resistant, the sensitivity of the lizard and the hedgehog to bungarotoxin lies somewhere between these two extremes.

The evolution of receptors

The findings of the immunological geneticists in Israel give us grounds for speculating about the way in which so many different forms of acetylcholine receptors have come to exist in our bodies. We have suggested that a messenger system was imperative if complex life forms, composed of a multitude of cells, each walled in by a membrane, were to survive in a changing or hostile environment. The earliest of these messenger systems, which has survived to the present day, appears to have been mediated by acetylcholine. By means of the acetylcholine receptor on the cell membrane, messages were decoded and converted into a brief flux of electrically charged ions, triggering off a specific response in the cell. The various functions that were coordinated by this system must have developed at different times during the evolution of the species, possibly millions of years apart.

Because of the long time lapse between these evolutionary events it would be anticipated that the chemical building blocks, from which these receptors were built, would vary slightly according to the availability of a supply of specific chemicals. The receptors in organs that had developed at an earlier stage in evolution would be likely to include different chemicals from those that appeared millions of years later. Provided the basic structure remained unchanged and it would still retain its sensitivity to acetylcholine, all these receptors would function in the same way, by opening the central channel. However, because of the inclusion of a different sequence of amino acids in their structure, their sensitivity to both acetylcholine, and to drugs that have a chemical structure, like acetylcholine, might vary.

The nerves were initially simple protrusions of nerve cells. These extensions allowed the cell to influence structures some distance away. Each cell had to perform two different functions in this increasingly complicated system. They had to produce a transmitter substance, so that their messages could be passed on down their nerves to an end organ; and they had to receive and interpret incoming messages received from cells in other parts of the brain.

Fulfilling these two different roles required two different types of receptor: one to control the production and release of acetylcholine and another to make them responsive to incoming commands from other cells. As a result the receptors developed an increasing complexity.

Those associated with outgoing messages retained their basic five-unit template but, because of slight differences in the makeup of the subunits, groups of receptors reacted in slightly different ways to the same transmitter. These receptors have not been as well studied as the receptors at the motor end plate, due to their inaccessibility. Those that have been examined show considerable pleomorphism, which is the taking on of two or more distinctly different forms in the same organisms.

At least eight different types of subunit have been identified in the brain, each responding to acetylcholine in a different manner. As a result the same transmitter, operating at different sites in the brain, can produce different responses. This variability has given the brain the flexibility required in order to meet the complexity of responses demanded by evolutionary development.

If this speculation is correct, we can begin to understand why different receptors, in the various organs of the body, respond in a different way to the same transmitter. It also explains why we may see a particular side effect with one muscle-relaxant drug in clinical practice but not with another, although they both block the action of acetylcholine at the neuromuscular junction and cause paralysis. It explains some of the unwanted side effects we see with some neuromuscular blocking drugs.

It has long been known that not all the muscles of the body are equally susceptible to curare-like drugs and that the onset of paralysis is slower in some muscles than in others. Research into this problem has produced several ingenious explanations to make them compatible with the competition theory. They have been based either on the time it takes for the blood to reach the muscle, or the delay in the drug reaching the receptor. These explanations have never been convincing, since autoradiographs demonstrate that a drug such as gallamine is more rapidly concentrated in cartilage, a tissue virtually devoid of any blood flow, than in muscle with its rich supply of blood vessels.

It is more likely that the difference in the onset time of paralysis in the various muscles is a reflection of minor differences in the structure of the binding domain of the receptor in different muscles. Not all the muscles of the body developed their physiological role at the same time in the evolutionary process. For example, the diaphragm became important in the breathing process only when animals were based on land and started to breathe air. It is probable that the acetylcholine receptors in the diaphragm developed at a later stage in the evolutionary process than those of the limbs. They would therefore be more likely to resemble receptors that developed in other muscles at about the same time, than those on the limb muscles. This would result in their having different binding capacities, which, in turn, would explain the differences observed in their responsiveness to relaxant drugs.

The sensitivity to curare-like drugs varies from one animal species to another. The reason for this is likely to be similar to the cause of the differences in the sensitivity to the snake venom that Fuchs and her co-workers found existed between the various animal species they studied. Although they are very different chemical compounds, both curare and bungarotoxin almost certainly act in much the same way at the receptor.

We now have a better idea of how curare-like drugs might act at the receptor. Only further experiments will tell us to what extent our speculations are right. If they are correct it offers a way of reconciling the competition theory with our findings in the isolated arm. If the curare-like drugs react at two adjacent but separate sites, in the same domain, on the receptor, then the two concepts are reconcilable. It explains why the experiments carried out on the rat's diaphragm in a water bath gave results that were not reproducible in the isolated arm. In the water bath the concentration of curare is kept constant while the amount of acetylcholine is slowly increased in each experiment.

In these experiments the critical factor controlling the amount of block is the water-bath concentration of drug, and the amount of block will depend on the ratio of curare to transmitter. However, in the isolated arm, where the concentration of drug in the blood falls rapidly to almost zero once the tourniquet is released, it is the effect of the binding of the blocking agent with the high-affinity site, close by acetylcholine recognition zone, that is the limiting factor in the recovery from the block. It reveals the intrinsic stickiness of the drug to this high-affinity site and its role in determining the rate and degree of recovery. It demonstrates that, under clinical conditions, it is the affinity between the drug and the binding site and not the concentration of drug in the blood that will control the duration of action of a muscle relaxant drug.

Only when the blood level remains higher than usual during recovery from a block, as it did in the case of Theresa Lopez, does it prevent the recovery of neuromuscular transmission. The proposition that curare-like muscle-relaxant drugs act simultaneously at two sites, at the receptor, the acetylcholine recognition site and a nearby binding site, explains the puzzle I was faced with in the patient with the Colles fracture.

CHAPTER FOURTEEN

New Drugs

New drugs

Until 1967, curare was the safest of the muscle relaxants available. The only drawback in clinical practice was that in high doses it caused a fall in blood pressure. This fall in blood pressure was known to Claude Bernard and formed the subject of one of his early demonstrations.

Bernard told his audience that he did not consider the fall was sufficient to have serious consequences. This was true in clinical practice until the advent of cardiac bypass surgery in the 1960s. The early subjects for cardiac surgery were invariably seriously ill. The mortality associated with the technique was so high that the risks were justified only in very ill patients. Most of them had narrowed, sticky valves that made them extremely sensitive to any drug that reduced the efficiency of a heart already weakened by disease.

Unfortunately, the fall in blood pressure produced by curare often reduced cardiac efficiency and compromised an already sick heart. The search for a curare-like drug that did not produce this effect resulted in the production of pancuronium, which was introduced into anaesthesia in 1967.

The idea behind pancuronium was to use the rigid steroid molecule as a scaffold on which to hang two acetylcholine molecules, rather as the rigid benzylisoquinoline molecule gave curare its rigidity. It was prompted by a demonstration by Professor Singh, an Indian pharmacologist, that a steroid that contained two choline-like groups incorporated within such a molecule, at almost 1 nm apart, had paralysing potential. Pancuronium was an immediate success. It did not affect the blood pressure and, although it had a little effect on the heart rate, this was much less than with other new drugs, such as gallamine. It soon became the drug of choice for sick patients undergoing cardiac surgery.

The ideal drug

In 1975, an international meeting was held at Westminster Hospital in London, bringing together pharmacologists, chemists and anaesthetists working in the field of muscle relaxants, in order to decide what properties were required in an ideal muscle-relaxant drug. The need for a drug that did not affect the cardiovascular system had been largely met with the introduction of pancuronium. The anxiety that remained was that of the reversibility of any residual paralysis.

High on their wish list of desirable properties was for a

drug that was very short acting and was completely and safely reversible within minutes of its administration. It was obvious from our findings about the vulnerability of the excretion of drugs to the depression of renal function that often occurred during anaesthesia that suggested that only a drug that was rapidly removed from the blood-stream, without the need for excretion by the kidneys, would meet this requirement.

It was this request that initiated a new line of drug development centred on Harvard University. Professor John Savarese and his colleagues set about developing a drug that would be rapidly destroyed by an enzyme in the blood so as to cause its blood concentration to fall within minutes of its administration. They argued that such a drug was likely to have a rapid onset-and-offset profile. After many years of disappointment, a drug that had a short life in the blood was produced. The puzzle was that it did not have the action profile that was anticipated. It was neither rapid in onset nor particularly short-acting. Its duration of action was only a little shorter than that of curare. However, because it was rapidly destroyed in the bloodstream it did prove to be easy to achieve complete safe reversal within minutes of its administration. It was mivacurium.

We now have an array of very safe muscle-relaxant drugs, which have very few side effects, even in high doses. Most are readily reversible. The flexibility offered by the array of new drugs with few, or limited, side effects is a tribute to the pharmaceutical wizardry of the chemists.

Today, with our better understanding of the way these drugs work and the conditions necessary for the safe, complete reversal of their paralytic effects, safe anaesthesia with good muscle relaxation can be produced, even in very sick patients. Recovery is invariably quick and complete, and unpleasant sequelae are rare. With muscle paralysis comes the need take over the patient's breathing. It was the realisation that ventilation had to be adjusted, according to each patient's particular requirement that was the last important step in achieving muscle relaxation, with a curare-like drug, with complete safety.

CHAPTER FIFTEEN

Lessons Learned

In 1942, when curare was first introduced into anaesthesia, no one could have foreseen the effect it would have, not only on anaesthesia but also on the practice of medicine.

If you are due to have an operation under general anaesthesia today, there is a five-to-one chance that you will receive a muscle-relaxant drug. It is probable, in the Western world, that the drug used is likely to be one or other of the new synthetic agents that have been developed to replace curare. It was inevitable that cheaper synthetic drugs would be produced that mimicked the paralysing properties of curare.

The present-day drugs are a vast improvement on the original curare preparation of Intocostrin that was produced from the crude specimen brought back from the jungles of South America by Richard Gill. As a result of knowledge gained through experience, the way they are used today is also very different from what was originally envisaged when curare was first introduced into medicine.

In 1942, Griffiths and Johnson used very small doses of intocostrin, far less than would have been required to cause total paralysis, in order to help produce muscle relaxation for abdominal surgery. They used the paralytic properties of the drug to augment the effect of the anaesthetic on the muscles of the abdomen. Before the advent of curare, it had been necessary to use deep planes of anaesthesia to provide sufficient muscle relaxation to allow surgeons to work inside the abdomen. In 1952, the eminent anaesthetist Harold Gillies pointed out that, 'Before curare, the deep planes of ether anaesthesia necessary for abdominal surgery invariably causes such severe depression of breathing and of the circulation that in inexpert hands, it was responsible for the deaths of many sick patients.'

According to Professor Francis Foldes, 'The administration of anaesthesia, in the days before muscle relaxants, was more of an art, mastered by relatively few, than a science that could be taught to many.'

Curare and the brain

In the USA, the early frightening experiences of Rovenstein in New York and Cullen in San Francisco, when they tried out the effect of samples of the experimental drug, gave rise to the idea that the dose of drug used should never be sufficiently large to stop the patient from breathing; spontaneous respiration had always to be maintained at all costs.

They taught that the deliberate paralysis of the muscles of ventilation constituted a physiological trespass. They emphasised that the muscle-relaxant drugs did not produce unconsciousness or reduce the effect of pain and that an adequate depth of anaesthesia should be attained before muscle paralysis was induced. The teaching was undoubtedly affected by the belief of eminent physiologists such as Seyle and Crile, that profound painful experiences could affect the brain and cause surgical 'shock' and death. Indeed, before 1950, surgical shock was not infrequently noted as a cause of death. Today, such a diagnosis is unthinkable. There is little doubt that it was the advent of curare that banished the concept of surgical shock.

Although the prevalent view in the USA was that curare should be used only to assist in relaxing the abdominal muscles, this attitude never found favour in Europe. In contrast to its use in the USA, curare was seen as part of anaesthesia, not merely an adjuvant, or enhancing agent. In the UK, the ability of muscle relaxants to reduce the overall anaesthetic requirements, whenever it was given, was seen as a justification of a much wider usage of the drug. It soon found favour in operations in which muscle relaxation was not a primary concern. The axiom 'If you can't anaesthetise them, paralyse them' came in to vogue. This idea brought undoubted advantages for both the anaesthetist and the patient. It meant it was quicker and easier for an anaesthetist to settle a difficult or sick patient, especially those who were very fat or seriously ill. As less anaesthetic was required, the patient made a quicker, safer and more comfortable recovery. The morbidity and mortality associated with general anaesthesia fell dramatically.

Within ten years of the introduction of curare into anaesthesia in the UK, the mortality associated with anaesthesia had fallen by about 30 per cent. Recovery from light anaesthesia was rapid and free of the prolonged after effects, such as vomiting and drowsiness, commonly seen before curare came to be used. The only drawbacks were an occasional difficulty in completely reversing the paralysis, leaving the patient weak and susceptible to respiratory problems in the postoperative period, and patients being aware of what was going on while they were believed to be asleep.

The British experience with very light anaesthesia and complete muscle paralysis, revealed that, provided the patient was asleep, the pain associated with surgical operations did not cause 'shock'. It is not that the muscle relaxant stopped pain, rather it did not register as a painful sensation if the patient was asleep. Even those patients who were vaguely aware that surgery was taking place seldom complained about the severity of the pain but rather the fear associated with the paralysis. As a result of the use of muscle-relaxant drugs, pain has come to be recognised as a subjective experience depending largely on the patient being conscious.

It has became apparent that a large dose of muscle

relaxant actually depresses the background information constantly flowing into the brain like background noise, which sets the level against which the brain prioritises incoming information. It dramatically reduces the incoming information from the sensory nerve endings in the joints and muscles that normally tell the brain about the position of our limbs and body. This background input occurs continuously, without our ever being aware of its happening, and it tends to reduce our awareness and to potentiate the effect of anaesthetics on consciousness.

Although it is not an anaesthetic, there are good grounds for believing it increases the effect of anaesthetic agents. One of the stranger side effects of this is that training, such as responding to a command or sensation, performed when an animal or a person is partially paralysed with a small dose of muscle relaxant is not reproducible if the same command or sensation is repeated when the subject is not paralysed. Clearly, the presence of even a minor degree of paralysis causes a change in the background activity of the brain, against which sensations and commands are interpreted. Although the muscle-relaxant drugs do not directly affect the brain, they do have an indirect action. This effect, which has been clearly demonstrated in animals, has been shown to occur in volunteers.

Artificial ventilation

At first there was little control over the amount of artificial ventilation given to curarised patients. Cecil

Gray, in Liverpool, taught that, provided the patient received more ventilation than he would if he breathed for himself, overventilation was all to the good. It took many years for it to be recognised that this overventilation, while it reduced the amount of anaesthetic required, did so by compromising the blood flow to the brain and causing potentially harmful changes in the blood.

It also imposed an extra burden on the heart. Anyone who has climbed to the top of a mountain without having taken time to become acclimatised will recognise the unpleasant effects brought on by overventilation. The reduced oxygen found in the air at high altitudes is associated with a fall in blood oxygen concentration. This causes the brain to speed up the rate and depth of ventilation in an effort to increase the oxygen taken up by the blood in the lungs. The overventilation reduces the amount of carbon dioxide in the blood, causing headache, nausea and a loss of appetite.

It has been found that in the unacclimatised individual, prolonged exposure to low carbon dioxide levels leads to a loss of memory. A similar loss of memory often occurred following an anaesthetic accompanied by overventilation. Gradually, as more and more studies on the physical changes in the blood caused by overventilation were reported and the disturbance it caused to various bodily processes became appreciated, the technique became modified and the artificial ventilation became more strictly controlled.

In the beginning the problem was that there was no easy way of measuring the amount of ventilation required by a patient when he was anaesthetised and paralysed. By the mid-1970s this problem was overcome with the development of carbon dioxide monitors. Today it is usual to measure constantly the amount and the effect of artificial ventilation so as to maintain the carbon dioxide level within narrow limits.

It was because of the need to control the ventilation of patients who were curarised that we became aware of the importance of aberrations in the carbon dioxide levels in the patient's blood and the effect it had on the acidity of the body fluids.

Until the practice of paralysing patients and taking over the responsibility for ventilating their lungs, we had failed to appreciate the precision with which our breathing is controlled by the brain so that not only do we take in sufficient oxygen to meet our needs but, more importantly, we keep our carbon dioxide level in the blood within a very narrow range. It is yet another example of the truism, proposed by Claude Bernard, that the maintenance of the internal environment in the body within very narrow parameters is essential for life. The main process by which we maintain the acid–base balance, essential for controlling the acidity of the body, is by altering the amount and depth of breathing. Ultimately we excrete any excess acid or base in our urine but this process takes many hours or days.

As carbon dioxide is the main acid-forming

component in our body, any sudden deviation from normal will upset the acid–base ratio and cause a change in the acidity of water inside the cells of our body. Too much acid or alkali in the cell water seriously interferes with its function. The brain functions rather like the helmsman of a boat, adjusting the tiller to correct any deviation from an allotted course when exposed to a stormy sea. Initially, the adjustments tend to be excessive but gradually they became smaller and smaller as a perfect course is achieved.

The appreciation of the danger associated with a major change in the ratio of acid to alkali led to an enormously important series of investigations in the 1960s and 1970s. These studies revealed the importance of ventilation in controlling the acid–base balance in the body. As a result of this knowledge new apparatus was designed to monitor ventilation and carbon dioxide levels. Today, we are able to ventilate sick patients safely for days, or weeks, in intensive-care units without upsetting their internal environment. This has allowed us to treat many previously fatal lung conditions.

Lessons from curarisation

Clearly we have come a long way from the first experimental use of the South American arrow poison. There can be no doubt that the widespread use of the muscle-relaxant drugs has changed the practice of anaesthesia. It has reduced surgical and anaesthetic mortality; it has speeded up the recovery from anaesthesia; and it has reduced the incidence of unpleasant side effects. In a review of the benefits resulting from the introduction of the muscle relaxants, Francis Foldes, whose career in anaesthesia started in the year that Griffith and Johnson published their report on the use of curare in patients, wrote:

> The advent of curare extended the boundaries of operability and patient safety. It made possible the explosive development of cardiovascular, intracranial and organ transplant surgery. For all practical purposes, the concept of 'inoperability' due to extremes of age, advanced disease or excessive trauma has been eliminated from medical terminology.

It has fulfilled the prediction that Gray and Halton made in a lecture to the Royal Society of Medicine in 1946, that the introduction of the muscle relaxants into medicine represented 'a milestone in anaesthesia', which they believed to be as important for anaesthesia as the introduction of asepsis had been for the progress of surgery.

Claude Bernard would have been delighted by the way the increase in our knowledge of the basic physiology of the body has been the fruit of our research into the way curare works and our exploitation of its effect in clinical practice. When he started his research into curare, he wrote in *Leçons sur les effets des substances toxiques et médicamenteuses* ('Lessons on the effects of toxic substances and of medicaments') that he believed that, by finding out how various poisons worked, we would learn about the way in which the various bodily systems function. The story of curare is a vindication of his prediction.

Epilogue

Our story began with the discovery of America and the tales that reached Europe of the deadly poison used by the natives to enhance the killing power of their arrows. This saga, which started over three hundred years ago, has led to an understanding of how curare the deadly arrow poison, produces paralysis and how its effect can be reversed.

Its introduction into medicine has resulted in a seismic change in anaesthetic practice. It was aptly described as a 'milestone in anaesthesia'. Not only did it reduce anaesthetic mortality by about 30 per cent, but it has led to the use of techniques that make possible a rapid and pleasant recovery after surgery. It has given us a better understanding of breathing and its effect on the acidity of body fluids; in new attitudes to prolonged artificial respiration that have led to the abandonment of the iron lung; and in a realisation of the importance of chemical transmitter substances in the control of our bodily functions.

It is the appreciation of the part played by these chemicals in our brain that has allowed new and exciting pharmacological developments, that offer us the prospect of drugs that will counter the effects of ageing on memory and balance, and provide a remedy for many intractable mental disabilities.

Chemical transmitters

There can be no doubt that Henry Dale and his team of pharmacologists at University College in London richly deserved the Nobel Prize and the accolades they received for putting chemical transmitters on the physiological map. At the time, they could not have envisaged the effect that their findings were to have on the practice of medicine. It was their convincing demonstrations in 1934–5 of the role of acetylcholine in the transmission of nervous signals that led to our present understanding of the way the brain functions.

Recent investigations have given us some insight into the complexity of the various chemical transmitter systems. Evolutionary changes that have taken place over millions of years have produced layer on layer of different, but closely integrated, chemical transmitter systems. Each layer of this complex interlocking system allows the brain to fulfil a part of its many functions. Unravelling these is like peeling the layers off an onion. The difficulty is compounded by not knowing the sequence in which the layers have evolved.

All the research suggests the primary chemical transmitter is acetylcholine and that it plays the key role in the system. It has been likened to the conductor of an orchestra who determines when other performers in this system come into play and with what strength they will respond.

It was thanks to researchers such as Gaskell and Langley in Cambridge, as well as Dale in London, that we came to recognise the important role of chemical transmitters in translating the brains commands into physiological responses. It is by means of these agents that the cells of the brain talk to each other so as to co-ordinate their activities. They are responsible for initiating a chain of events, in response to incoming information, that causes the instructions emanating from the brain to travel down the nerves to the organs it controls.

It is the chemical transmitters, released at nerve endings, that produce the reaction required in the target organ. It has been the demonstration of the central part played by these chemicals in the control of bodily functions, and in the complex feedback mechanisms that they modulate, that has been responsible for a huge leap forward in our understanding of the way the brain works.

We have come to recognise that acetylcholine is only one of a stable of chemicals used by the brain to communicate with the rest of the body. It was fortunate that it was the first transmitter whose role was established, as it is probable that it, or a substance very like it, was the first of the series of chemicals that have evolved to meet the increasingly sophisticated demands of more complex forms of life. It is the key chemical in what is now known to be a complex interactive control system. Because of its evolutionary primacy, acetylcholine has come to be central to many of the brain's activities. It directly transmits the brain's instructions to those organs concerned with the basic survival of the body: the heart, the gut, the digestive system and the muscles.

Elsewhere, it causes a different response by initiating the release of secondary transmitter chemicals. In this process it acts like an army runner passing on the orders of the general in the high command to the various units in the field. These secondary transmitters, such as adrenaline and noradrenaline, are mainly associated with the changes required to fulfil more aggressive survival responses, such as increasing the heart rate and blood pressure in preparation for fighting or fleeing, and reducing the bowel activity so that blood can be diverted to meet more pressing, immediate needs.

In other areas of the brain, acetylcholine may cause the release of the neurotransmitter serotonin, which causes emotional changes. It causes an increased awareness and an elevation in mood. It plays a major role in states such as hunger, hallucinations and depression. In other areas it increases the dopamine level, causing the suppression of coarse muscle movements. Without the emotional responses associated with these secondary transmitters, the higher levels of social behaviour and group interactions, that we take for granted, would be impossible.

Other transmitters such as GABA (gamma ameno hydroxy butyric acid) produce a sleep-like state and relaxation, while endorphins, which act on a particular small area of the brain, act as natural opiates causing euphoria and a reduction in pain.

A major problem in understanding the part played by the various transmitters is compounded by the receptors. Small changes in their chemical composition have allowed the same transmitter to produce a different response. Although the action of acetylcholine is always the same – in that it opens the central channel in the receptor – the response it produces in the body differs according to the type of receptor it activates. By exploiting these subtle differences it is expected that it will be possible to block isolated effects of a transmitter at some receptors without affecting others.

The advances in therapeutic medicine associated with these developments have led to the development of drugs such as beta blockers, which selectively antagonise some of the effects of adrenaline but not others; antidepressant drugs such as Prozac and amitriptyline, which increase the brain levels of serotonin; drugs to treat Parkinson's disease; anti-emetic and anti-diarrhoea agents; drugs to treat asthma; drugs to reduce appetite; and others to control manic states. Many of these drugs have changed the pattern of disease and have opened the possibility of patients, who had been confined to institutions, living a reasonably normal life in the community. The next exciting step is to produce drugs that act only on one particular set of receptors without affecting others, the so-called 'magic bullets'. These would target a local deficiency of a specific transmitter such as the one believed to be responsible for many of the mental changes associated with ageing.

Although our story concerns curare, it is not the only herb affecting the chemical transmitter system that has been used by man. We can now understand why Chinese peasants chew the leaves of *Ephedra vulgaris* and the natives of South America cocoa beans. Both contain chemicals that potentate the action of adrenaline and reduce the pangs of hunger. Betel nut and fly agaric have been chewed for hundreds of years by the natives of Asia to overcome thirst and lethargy. They contain chemicals that block the actions of acetylcholine that cause salivation, and they can produce increased awareness and mood changes. Belladonna berries were used by Roman women to achieve the 'doe-eyed' look. We now know this effect is due to its blocking action on acetylcholine in the eye. The use of tobacco, a herb that stimulates some of the acetylcholine receptors, is a relatively recent addition to this list. The realisation of the way many of these native remedies work has been a stimulus to the pharmacological exploration of herbal medicine. It is probable that we still have much to learn from chemicals that abound in the wild.

A visitor to the Amazonian jungle today can still find villagers living in much the same way as they did in the days of Charles Waterton and Richard Gill. They still hunt their food and live off the produce of the forests, but today they are more likely to use guns than curare darts. Seeing their unchanged, primitive lifestyle, it is difficult to imagine that it was from their domain that the world received news of the much-feared poison that was, one day, to help unlock the story of the workings of our brain and to bring us remedies for so many diseases common in the developed world.

This story, which began with the exploration of the New World, has not yet ended. It has resulted in profound changes in our understanding of the way our body works; it has opened a new era in anaesthesia and medicine; it is an ongoing saga that started over three-hundred years ago

It is the story of the South American arrow poison: curare.

Further Reading and References

It seemed to me that any attempt to reference the source of all the material in this book would produce a very long list, which would do little to enlighten the reader wishing to dip more deeply into the subjects covered here. I have therefore restricted the list to other sources of the important information on the stories in the book, and to books and papers that contain a more detailed account on the subject.

Details of the experiments carried out in the Research Department of the Magill Department of Anaesthetics at the Chelsea and Westminster Hospital and referred to in this book can be found in *Neuromuscular Block* (Butterworth Heinemann, 1996), by this author.

Bibliography

Bernard, Claude, (1865) *An introduction to the Study of Experimental Medicine* (tr. H. Copley-Greene) (Macmillan & Co.)

Bernard, Claude, Bailliere et fils, *Leçons sur les effets des substances toxiques et médicamenteuses.* ('Lessons on the effects of toxic substances and of medicaments') (Paris; tr. Thornton) (Williams & Wilkins Press, Baltimore, 1961)

Bowman, W. C., *Pharmacology of Neuromuscular Function* (Wright, 1990)

Burnap, T. K., and Little, D. M., *The Flying Death* (Little, Brown & Co., Boston, 1968)

Claude Bernard and his Place in the History of Ideas. R. (University of Nebraska Press, 1960)

Columbus, Ferdinand, *The Life of Admiral Christopher Columbus* (Rutgers Press, New Brunswick, 1958)

Dale, H. H., *Adventures in Physiology with Excursions into Pharmacology* (Pergamon Press, 1953)

Dor-Ner, Zvi, *Columbus and the Age of Discovery* (HarperCollins, 1991)

Feldberg, W. S., *Biographical Memoirs of Fellows of Royal Society* (1970)

Feldman, S.A., *Muscle Relaxants* (W.B. Saunders, 1979, 1983).

Feldman, S. A., *Neuromuscular Block* (Butterworth Heinemann, London, 1996)

Feldman, S. A., Scurr, C. F., and Paton, W., *Drugs in Anaesthesia* (Edward Arnold, 1987)

Gill, Richard C., *White Waters Black Magic* (Henry Holt & Co., New York, 1940)

Greenfield, Susan A., *The Private Life of the Brain* (Allen Lane, 2000)

Hakluyt, A., 'Relating to the Second Voyage to Guinea', excerpted in K. Bryn Thomas, *Curare, its History and Usage* (Pitman, London, 1964)

Hawkins, C., *The Works of Sir Benjamin Brodie* (London, 1865)

Innes, Hammond, *The Conquistadors* (Collins, 1969)

Koelle, G. B., *Reflections of the Pioneers of Neurohumoral Transmission* (Springer-Verlag, 1985)

Leonardo, R. A., *Lives of Master Surgeons* (Forben Press, New York, 1948)

McIntyre, A. R., *Curare: Its History, Nature and Clinical Uses* (University of Chicago Press, 1947)

Physiology of Medicine, Nobel Lectures (Elsevier Publishing Co., 1965)

Porter, R., *The Greatest Benefit to Mankind* (HarperCollins, 1997)

Raleigh, Sir Walter, *The Discovery of the Large, Rich and Beautiful Empire of Guinea* (London, 1956)

Roberts, M. B., *Nothing is Without Poison* (Chinese University Press, 2002)

Robin, E. Debs (ed.), *Claude Bernard and the Internal Environment*, proceedings of a symposium held at Stanford University, USA, 1978 (Marcel Dekker, New York 1979)

Smith, P., *Arrows of Mercy* (Doubleday, New York, 1969)

Thomas, K. Bryn, *Curare, its History and Usage* (Pitman, London)

Turner, J., *Spice: The History of Temptation* (London, 2004)

Walker, Gabrielle, *Snowball Earth* (Bloomsbury Press, 2003)

Waterton, C., *Wanderings in South America* (Macmillan & Co. 1879; republished by Echo Library, 2005)

Papers and abstracts

Agoston, S., Feldman, S. A., and Miller, R. D., 'Plasma pancuronium concentrations following the depression of twitch height in the isolated arm, following bolus injection and continuous infusion', *Anesthesiology*, 1979

Barchan, S., et al., 'How the Mongoose can fight the Snake: the binding site of the mongoose receptor', *Proceedings of the National Academy of Science*, Vol. 7717, 1992

Beecher, H. K., and Todd, D. P., 'Study of Deaths Associated with Anesthesia and Surgery in 10 American Institutions between 1948–1952', *Annals of Surgery*, 1956

Bennett, A. E., 'Clinical Investigations with Curare in Organic Neurologic Disorders', *Journal of the American Medical Association*, Vol. 202, 1940

Bennett, A. E., 'Preventing Traumatic Complications of Convulsive Shock Therapy', *Journal of the American Medical Association*, Vol. 144, 1940

Birks, R. J., and McIntosh, F. C., 'Acetylcholine metabolism at nerve endings', *British Medical Bulletin*, Vol. 13, 1957

Bowman, W. C., 'On pre- and postjunctional cholinoreceptos at the neuromuscular junction', *Anesthesia & Analgesia*, Vol. 59, 1980

Brodie, B. C., 'Experiments and Observations on the different Modes in which Death is produced by certain vegetable Poisons', Society for Promoting the Knowledge of Animal Chemistry, *Annals of the Royal Society*, 1811

Brodie, B. C., 'Further Experiments and Observations on the Action of Poisons on the Animal System', communicated to the Society for the Improvement of Animal Chemistry and the Royal Society, *Annals of the Royal Society*, February 1812

Cohen, E. N., Hood, N., and Golling, R., 'Use of whole body autoradiography to determine the uptake and distribution of d-tubocurarine H3', *Anesthesiology*, Vol. 28, 1968

Cullen, S. C., 'Curare in Anaesthesia', *Surgery*, Vol. 18, 1945

Dale, H. H., Feldberg, W., and Vogt, M., 'Release of acetylcholine at voluntary motor nerve endings', *Journal of Physiology*, Vol. 86, 1936 del Castillo, J., and Katz, B., 'A study of curare with an electrical micro method', *Proceedings of the Royal Society of Biology*, Vol. 146, 1954

Feldman, S. A., and Tyrrell, M. F., 'A New Theory of the Termination of Action of Muscle Relaxants' *Proceedings of the Royal Society of Medicine*, Vol. 63, 1970

Ferguson, A., 'A forgotten pharmacologist – Abbé Felice Fontana', *Anaesthesia*, 2004

Fick, G. R., 'Henry Dale's involvement in the verification and acceptance of the theory of neurochemical transmission: A lady in waiting', *Journal of the History of Medicine*, Vol. 42, 1987

Fuchs, S., et al., 'Molecular Evolution of the Binding Site of the Acetylcholine Receptor', *Annals of New York Academy of Sciences*, Vol. 126, 1994

Galindo, A., 'Non-depolarizing neuromuscular block', *Journal of Pharmacology and Experimental Therapeutics*, Vol. 178, 1971.

Gray, T. C., and Halton, J. A., 'A Milestone in Anaesthesia', *Proceedings of the Royal Society of Medicine*, 1946

Griffith, H. R., and Johnson, E., 'The use of curare in general anesthesia', *Anesthesiology*, Vol. 418, 1942

Hatcher, J., 'Spencer Wells and his famous forceps', *Midwives Chronicle*, Vol. 86, 1973

Hill, A. V., 'The mode of action of nicotine and curari, determined by the form of the contraction curve and the temperature coefficient', *Journal of Physiology*, Vol. 39, 1909

Hodgkin, A. L., and Katz, B., 'The effect of temperature on the electrical activity of the giant axon of the squid', *Journal of Physiology*, Vol. 109, 1949

Hunt, R., and Traveau, R., 'On the pharmacological actions of certain choline derivatives', *British Medical Journal*, 1906

Jenkinson, D. H., 'The antagonism between tubocurarine and substances that depolarise the end plate', *Journal of Physiology*, Vol. 152, 1960

Katz, B., 'Quantal mechanism of neural transmitter release', *Science*, Vol. 173, 1971

King, H. F., 'Curare', *Nature*, Vol. 135, 1935

Langley, J. N., 'On nerve endings and on special excitable substances in cells', *Proceedings of the Royal Society*, Vol. 78, 1906

Loewi, O., 'Über humorale ubertragbarkeit der herenvenwirkung', *Pflugers Archives*, Vol. 189, 1921

Matteo, R. S., Spector, S., and Horowitz, P. E., 'Relation of serum d-tubocurarine concentration to neuromuscular block in man', *Anesthesiology*, Vol. 41, 1974

Neher, E., and Sackman, B., 'Single channel currents recorded from the membrane of the denervated frog muscle fibre', *Nature*, Vol. 260, 197

Paton, W. D. M., and Zaimis, E. J., 'The pharmacological actions of polymethylene bis-methyl ammonium salts', *British Journal of Pharmacology and Chemotherapy*, Vol. 4, 1949

Robertson, J. D., 'Electron microscopy of the motor end-plate and the neuromuscular spindle', *American Journal of Physiology*, Vol. 39, 1960

Saverese, J. J., Ali. H. M., and Basts, S. J., 'The clinical neuromuscular pharmacology of mivacurium chloride', *Anesthesiology*, Vol. 68, 1988

Sayers, L. A., 'Cases of Traumatic Tetanus', *New York Journal of Medicine*, 1858

Shepard, J. A., 'Spencer Wells – surgeon RN', *Proceedings of Royal Society of Medicine*, Vol. 63, 1907

Standaert, F. G., and Adams, J. E., 'The actions of succinylcholine on mammalian nerve terminals', *Journal of Pharmacology and Experimental Therapeutics*, Vol. 149, 1965

Sven, J., 'Felix Fontana', thesis submitted to University of Leiden, 1942

Sven, J., 'Neuromuscular blocking drugs: common interactions in daily clinical practice', *University of Leiden Press*, 1941

Wells, Spencer T., 'Three Cases of Tetanus in which "Woorara" was used', *Proceedings of the Medical and Chirurgical Society of London*. Vol. 3, 1859